Empirical Analysis of
Natural Gas Markets

Empirical Analysis of Natural Gas Markets

Editors

David A. Anderson
Shigeyuki Hamori

MDPI • Basel • Beijing • Wuhan • Barcelona • Belgrade • Manchester • Tokyo • Cluj • Tianjin

Editors
David A. Anderson
Dept. of Economics & Finance
Centre College
USA

Shigeyuki Hamori
Graduate School of Economics
Kobe University
Japan

Editorial Office
MDPI
St. Alban-Anlage 66
4052 Basel, Switzerland

This is a reprint of articles from the Special Issue published online in the open access journal *Energies* (ISSN 1996-1073) (available at: https://www.mdpi.com/journal/energies/special_issues/Empirical_Analysis_Natural_Gas_Market).

For citation purposes, cite each article independently as indicated on the article page online and as indicated below:

LastName, A.A.; LastName, B.B.; LastName, C.C. Article Title. *Journal Name* **Year**, *Article Number*, Page Range.

ISBN 978-3-03943-136-6 (Hbk)
ISBN 978-3-03943-137-3 (PDF)

© 2020 by the authors. Articles in this book are Open Access and distributed under the Creative Commons Attribution (CC BY) license, which allows users to download, copy and build upon published articles, as long as the author and publisher are properly credited, which ensures maximum dissemination and a wider impact of our publications.

The book as a whole is distributed by MDPI under the terms and conditions of the Creative Commons license CC BY-NC-ND.

Contents

About the Editors vii

Preface to "Empirical Analysis of Natural Gas Markets" ix

David A. Anderson
Natural Gas Transmission Pipelines: Risks and Remedies for Host Communities
Reprinted from: *Energies* **2020**, *13*, 1873, doi:10.3390/en13081873 1

Jin Shang and Shigeyuki Hamori
The Response of US Macroeconomic Aggregates to Price Shocks in Crude Oil vs. Natural Gas
Reprinted from: *Energies* **2020**, *13*, 2603, doi:10.3390/en13102603 11

Yulian Zhang and Shigeyuki Hamori
Forecasting Crude Oil Market Crashes Using Machine Learning Technologies
Reprinted from: *Energies* **2020**, *13*, 2440, doi:10.3390/en13102440 29

Wenting Zhang and Shigeyuki Hamori
Do Machine Learning Techniques and Dynamic Methods Help Forecast US Natural Gas Crises?
Reprinted from: *Energies* **2020**, *13*, 2371, doi:10.3390/en13092371 43

Tiantian Liu, Xie He, Tadahiro Nakajima and Shigeyuki Hamori
Influence of Fluctuations in Fossil Fuel Commodities on Electricity Markets: Evidence from Spot and Futures Markets in Europe
Reprinted from: *Energies* **2020**, *13*, 1900, doi:10.3390/en13081900 65

Tadahiro Nakajima and Yuki Toyoshima
Examination of the Spillover Effects among Natural Gas and Wholesale Electricity Markets Using Their Futures with Different Maturities and Spot Prices
Reprinted from: *Energies* **2020**, *13*, 1533, doi:10.3390/en13071533 85

Guizhou Liu and Shigeyuki Hamori
Can One Reinforce Investments in Renewable Energy Stock Indices with the ESG Index?
Reprinted from: *Energies* **2020**, *13*, 1179, doi:10.3390/en13051179 99

Wenting Zhang, Xie He, Tadahiro Nakajima and Shigeyuki Hamori
How Does the Spillover among Natural Gas, Crude Oil, and Electricity Utility Stocks Change over Time? Evidence from North America and Europe
Reprinted from: *Energies* **2020**, *13*, 727, doi:10.3390/en13030727 119

Yijin He, Tadahiro Nakajima and Shigeyuki Hamori
Connectedness Between Natural Gas Price and BRICS Exchange Rates: Evidence from Time and Frequency Domains
Reprinted from: *Energies* **2019**, *12*, 3970, doi:10.3390/en12203970 145

Tadahiro Nakajima and Yuki Toyoshima
Measurement of Connectedness and Frequency Dynamics in Global Natural Gas Markets
Reprinted from: *Energies* **2019**, *12*, 3927, doi:10.3390/en12203927 173

About the Editors

David A. Anderson is the Paul G. Blazer Professor of Economics at Centre College. He received his BA from the University of Michigan and his MA and PhD from Duke University. Dr. Anderson's research focuses on the economics of the environment, law, crime, and public policy. His other books include *Environmental Economics and Natural Resource Management*, *Survey of Economics*, and *The Cost of Crime*.

Shigeyuki Hamori is a professor of Economics at the Graduate School of Economics, Kobe University, Japan. He completed his PhD in Economics from Duke University, United States. His research interests include applied time-series analysis, empirical finance, data science, and international finance.

Preface to "Empirical Analysis of Natural Gas Markets"

Recent developments in the natural gas industry warrant new analyses of related issues and markets. Abundant supplies of natural gas unearthed by hydraulic fracturing have altered the landscape for energy economics. Environmental, social, and governance (ESG) investments have accelerated the shift away from coal as the dominant source of electricity, in part because natural gas is the cleanest burning fossil fuel. The processing and liquefaction of natural gas remove most of its impurities, and compared to petroleum and coal combustion, natural gas combustion releases relatively little CO2 and NOX, among other pollutants. Its low environmental impact and reduced volume make liquefied natural gas (LNG) a popular source of energy during this time of transition between traditional fuels and newer options. Broad availability furthers the appeal of LNG. Unlike oil, whose sources are concentrated geographically, natural gas is extracted on six continents. In the United States, the shale gas revolution has made natural gas a game changer. Due to its many sources, even countries that import LNG can limit their supply-side risk with strategies that diversify their suppliers. In this book, we focus on empirical analyses of the natural gas market and its growing relevance worldwide.

David A. Anderson, Shigeyuki Hamori
Editors

Article

Natural Gas Transmission Pipelines: Risks and Remedies for Host Communities

David A. Anderson

Department of Economics and Finance, Centre College, 600 W. Walnut St., Danville, KY 40422, USA; anton42@adelphia.net

Received: 23 March 2020; Accepted: 10 April 2020; Published: 12 April 2020

Abstract: Transmission pipelines deliver natural gas to consumers around the world for the production of heat, electricity, and organic chemicals. In the United States, 2.56 million miles (4.12 million km) of pipelines carry natural gas to more than 75 million customers. With the benefits of pipelines come the risks to health and property posed by leaks and explosions. Proposals for new and recommissioned pipelines challenge host communities with uncertainty and difficult decisions about risk management. The appropriate community response depends on the risk level, the potential cost, and the prospect for compensation in the event of an incident. This article provides information on the risks and expected costs of pipeline leaks and explosions in the United States, including the incident rates, risk factors, and magnitude of harm. Although aggregated data on pipeline incidents are available, broadly inclusive data do not serve the needs of communities that must make critical decisions about hosting a pipeline for natural gas transmission. This article breaks down the data relevant to such communities and omits incidents that occurred offshore or as part of gas gathering or local distribution. The article then explains possible approaches to risk management relevant to communities, pipeline companies, and policymakers.

Keywords: natural gas; transmission; pipelines; external cost; health; property damage; bodily injury; uncertainty; insurance

1. Introduction

Natural gas pipelines extend through every state in North America to connect producers, distributors, and customers. Proposals to construct, expand, or repurpose pipelines often lead to contention over risks to host communities. Examples include recent debates over the Mountain Valley, Atlantic Coast, and Tennessee Gas pipelines. Some communities have enacted policies to deter natural gas pipelines [1], while others have welcomed them [2]. Decisions about pipeline construction and regulation are often made with scant information about the risks and costs for host communities. If well informed, prospective host communities can weigh the risks associated with natural gas transmission against the long-term benefits. This article provides information to assist communities and pipeline operators with the appropriate cost–benefit analysis, and offers possible remedies for the problems communities face regarding risk spreading and uncertainty.

In 2019, pipelines supported annual expenditures of almost $150 billion on natural gas in the United States for the production of heat, electricity, plastics, fertilizers, pharmaceuticals, fabrics, and organic chemicals, among other uses [3]. Figure 1 shows the steadily increasing use of natural gas in the United States. Figure 2 shows the percentage of natural gas used for each purpose. With the benefits of natural gas transmission come the threats of damage to life and property. After the construction phase of pipelines, the external costs stem largely from leaks or the combustion of toxic loads and the resulting damage to property, health, and the environment.

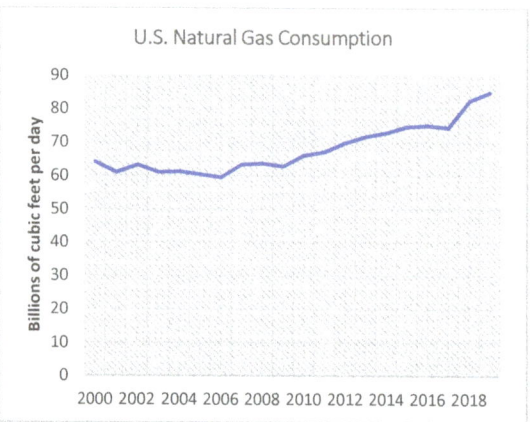

Figure 1. U.S. natural gas consumption, 2000–2019. Data source: U.S. Energy Information Administration.

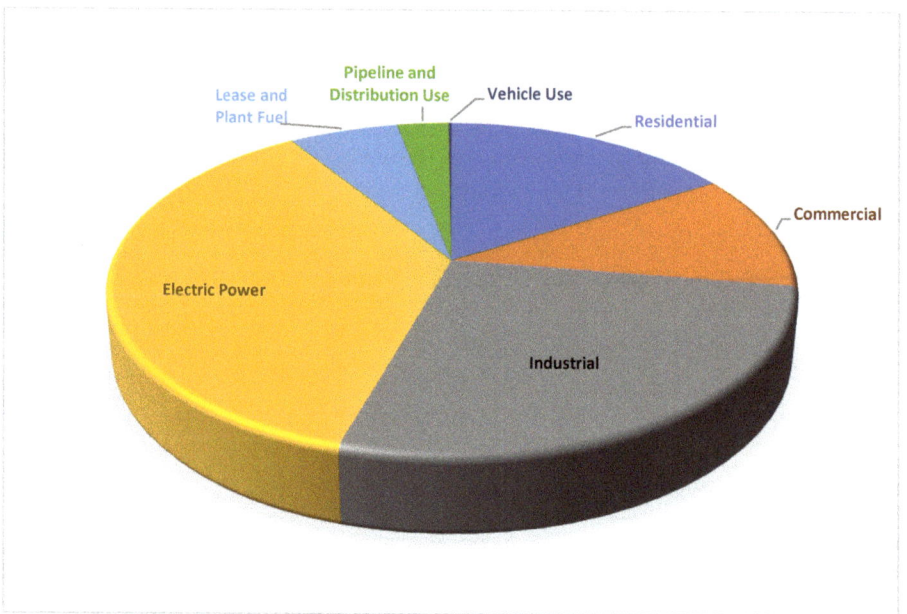

Figure 2. U.S. natural gas consumption by sector, 2019. Data source: U.S. Energy Information Administration.

The Pipeline and Hazardous Materials Safety Administration (PHMSA) reports a total of 12,316 natural gas, hazardous liquids, and liquefied natural gas pipeline incidents between 2000 and 2019 [4]. The repercussions included 309 deaths, 1232 injuries, and $10.96 billion in property damage. These figures are accessible, yet a majority of the underlying incidents are irrelevant to communities that might host a natural gas transmission pipeline. Some of the incidents occurred offshore, some involve more volatile substances than natural gas, and some occurred in gas gathering and distribution operations that stem from a different set of decisions than transmission pipelines.

This article focuses on the information relevant to prospective host communities for natural gas pipelines. Using data paired down to include only onshore natural gas transmission pipeline incidents, the article provides incident rates and estimated costs of bodily injury, lost life, and property damage. Regression analysis provides further insights into expected damage costs based on community and pipeline characteristics. The article also discusses approaches to risk management for communities that could be applied in any country.

Section 2 of this article provides a review of the related literature. Section 3 explains the methods used to establish the dataset and estimate the regression coefficients. Section 4 reports the results. Section 5 discusses implications and possible remedies for host communities' exposure to risk and uncertainty. Section 6 concludes the paper.

2. Literature Review

The previous literature on the costs of pipelines to host communities focuses largely on the effects of pipelines on property values [5]. Reductions in property values near pipelines reveal perceptions of the risks of leaks and explosions. If consumers were informed, rational, and risk-neutral, the loss in property values would accurately reflect the expected cost of such incidents, and it would be redundant to add pipeline-related decreases in property values to the costs of property damage, injuries, and deaths when calculating the total cost of pipeline incidents. If consumers have imperfect information, the effect of pipelines on property values is not an accurate measure of the expected cost. Residents who perceive no risk of pipeline incidents are willing to pay the same amount for a home regardless to its proximity to a pipeline.

The findings on pipelines' effects on property values are inconclusive. Studies by McElveen et al. [6] and Integra Realty Resources [7] are among those suggesting that pipelines have no significant influence on property values. In contrast, Simons et al. [8] and Hanson et al. [9] estimate that major incidents involving oil and gas pipelines lower property values by 10.9%–12.6% and 4.65%, respectively. Kielisch [10] provides evidence from realtors, homeowners, real estate appraisers, and land sale analysis that natural gas pipelines can lower property values significantly, and in some cases by as much as 39%. Herrnstadt and Sweeney [11] point out that accurate information on pipeline risks would allow people to respond with appropriate safety plans. Another benefit of the present research is that it provides information with which homebuyers can make better decisions about their willingness to pay for homes near pipelines.

Another body of research presents models of risk assessment for pipelines. That research allows operators to fine-tune their risk estimates based on situation-specific characteristics such as the density and pressure of gas within the pipeline [12–14]. The present research incorporates broader community characteristics such as mean income and population density, along with the age of the pipeline, as determinants of the cost of an accident. The latter determinants are relatively constant and readily available to communities considering the prospect of a new pipeline.

Economists must reluctantly place a value on human life to inform decisions about tradeoffs between money and lives, including decisions about safety regulations, environmental policies, and pipelines. The existing literature addresses the value of unidentified or "statistical" lives such as the lives that could be lost by a community hosting a pipeline. We know that statistical lives have finite value because communities make decisions that have finite benefits and involve risk to life. By allowing people to drive cars, cross streets, operate farm machinery, smoke, and use natural gas, it is inevitable that deaths will result. If the value of a statistical life were infinite, none of these activities would be acceptable. Estimates of the value of a statistical life come from real-world tradeoffs people make between money and risks of death as revealed in labor markets among other settings. In a recent synthesis of the available research, Viscusi [15] estimated that the bias-corrected mean value of a statistical life is $10.45 million. Several U.S. agencies apply similar estimates, including the Occupational Safety and Health Administration, the Food and Drug Administration, and the Environmental Protection Agency. A related vein of literature exists for the value of bodily injury. Viscusi and Aldy [16] provide a summary of 24 relevant studies of the value of a statistical injury, the mean of which is $90,697. These values for a statistical life and a statistical injury are applied to deaths and injuries in the present study.

3. Data and Methods

Data on natural gas pipeline incidents are available from the Pipeline and Hazardous Materials Safety Administration (PHMSA), a division of the U.S. Department of Transportation [17]. The PHMSA dataset offers information on every reported natural gas pipeline incident in the United States, including the location of the incident, the cost of property damage, the number of injuries and deaths, and the age of the pipeline. The PHMSA requires that incidents be reported if they cause a death or in-patient hospitalization; at least $50,000 in property damage excluding lost gas; the unintentional loss of at least 3 million cubic feet of gas; an emergency shutdown of an underground natural gas storage facility; or an event that is significant in the judgment of the operator, even if it does not meet the other criteria [18].

Natural gas pipeline incidents involve both explicit and implicit costs. The explicit costs include the costs of public and private property damage and emergency responses, all of which are reported to the PHMSA. The implicit costs are the costs of injuries and lost life, estimated by multiplying the number of injuries and deaths by the value of each type of occurrence drawn from the literature on the value of a statistical injury and life [15,16].

For the regression analysis, those pipeline data were paired at the zip-code level with information from the U.S. Bureau of the Census on population density, and information from the Statistics of Income Division of the U.S. Internal Revenue Service on income, real estate taxes, and the percentage of tax returns that are farm tax returns. The population data come from the 2010 census, conducted halfway through the 2000–2019 time period being studied. The tax data came from 2017, the most recent year for which complete data were available. Table 1 provides variable definitions for the dataset.

Table 1. Variable definitions and summary statistics.

Variable	Definition	Mean (n = 1625)	Standard Deviation
Cost of incident	Total incident cost, including property damage, bodily injury, and loss of life, in 2020 USD.	$1,483,120	$19,600,000
Property damage	Inflation-adjusted total property damage	$1,205,772	$17,100,000
Population density	Population per square mile in that zip code area	624.80	3788.58
Mean income	Mean income in that zip code area (USD)	$59,108.35	$27,333.19
East	Regional dummy variable	0.12	0.33
Midwest	Regional dummy variable	0.26	0.44
West	Regional dummy variable	0.14	0.34
South	Regional dummy variable	0.49	0.50
Pipeline age	Years since pipeline's installation	38.86	23.42
Real estate taxes	Real estate taxes collected per square mile in that zip code area in 2017 (1000s of USD)	$333,934	$1,266,637
Percent farms	Percent of tax returns that are farm tax returns	5.31	6.88

The selected variables represent location characteristics that could influence the consequences of a pipeline incident. In related regression analysis of property damage from hazardous liquid pipeline incidents, Restrepo et al. [19] use a dummy variable for high-consequence areas, which include areas with high population density. The present research uses population density among other values that similarly affect incident cost. The specification was subject to the availability of data. It would be ideal to have measures of the population density and the value of real estate within close proximity of the pipeline. Actual data are available at the zip code level, which is not always limited to areas in close range of the pipeline. The specification was adjusted in response to empirical findings on the contribution of particular variables, as discussed further below.

The population density, mean income, and real estate taxes per square mile could each influence damage costs positively or negatively. A higher population density could increase the likelihood of a leak or explosion being near buildings and people. At the same time, areas with high population densities can have stricter requirements for pipe strength, stress levels, or monitoring, which decrease the likelihood of a high-cost incident [20]. Higher mean income similarly increases the likelihood that an incident of any particular scale would cause costly damage, but correlates with greater protections against major incidents. For example, Pless [21] reports that the U.S. state with the lowest mean income, Mississippi, had 50.7 inspection person days per 1000 miles (1609 km) of natural gas transmission pipeline in 2009, whereas the U.S. state with the highest mean income, Massachusetts, had 764.3. Controlling for population density and mean income, having higher real estate taxes per square mile is hypothesized to have a positive influence on the cost of property damage because, for any given tax rate, it rises with property values.

Incidents along gathering and distribution lines are not included in this research, because they result from a different decision-making process than transmission lines. The risk of incidents along transmission lines is an inherent aspect of playing host for the natural gas industry as it brings its product to distant markets. In contrast, distribution lines are the result of consumers in each municipality deciding to use natural gas as fuel. Further, many of the incidents in the distribution pipeline category occur at customers' homes and businesses. Gathering lines are in a distinct category as well. They are part of the natural gas production process and serve the purpose of bringing fuel from the extraction site to a central collection site. Offshore pipeline incidents are not included, because they are not related to the issue of communities hosting transmission pipelines.

The primary equation used to estimate the determinants of pipeline incident costs is

$$\ln Cost_i = \alpha_0 + \beta_1 Region_i + \beta_2 Population\ Density_i + \beta_3 \ln Mean\ Income_i + \beta_4 \ln Pipeline\ Age_i + \beta_5 \ln Real\ Estate\ Taxes_i + \epsilon_i$$

Region is a vector of the *East*, *Midwest*, and *South* dummy variables. The *West* dummy variable is omitted to avoid multicollinearity. Zip codes starting with 0–2 are in the East, those starting with 4–6 are in the Midwest, those starting with 8 or 9 are in the West, and those starting with 3 or 7 are in the South. The dummy variable *Midwest* is used instead of *North* because the observation level is zip-code areas, which are numbered from east to west. Zip-code areas starting with 8 and 9 run from the northern border to the southern border of the United States. It is, therefore, more practical to delineate the Midwest and West regions. The empirical investigation included several variations of this equation to assure the robustness of the findings.

4. Results

Looking only at the data on onshore natural gas transmission pipelines, between 2000 and 2019, there were 1846 incidents, 49 deaths, 173 injuries, and $1.7 billion in property damage. Of the 12,316 total pipeline incidents reported in the introduction, only 15% were along onshore transmission pipelines, which shows the importance of breaking out this category of incidents. Table 2 separates these figures by region for the most recent decade. The West had the most deaths, the most injuries, and the most property damage, despite having the second-lowest number of incidents. The coefficients on the regional dummy variables in the regression findings below support this finding and provide further insights into regional differences.

Table 2. Incidents and damage by region, 2010–2019.

	Incidents	Property Damage (2020 USD)	Injuries	Deaths
East	118	$148,302,230	2	0
Midwest	252	$173,782,636	17	6
West	151	$748,293,105	67	10
South	486	$254,581,542	20	9

There are about 300,000 miles (482,803 km) of onshore natural gas transmission pipelines in the United States, and there were 115 incidents in 2019. Table 3 shows the number of incidents per 10,000 miles (16,093 km) of these pipelines over the past 20 years. The numbers are notably consistent, with a mean of 3.11 and a standard deviation of 0.596. This indicates the relative predictability of incidents on a national scale and the inability of current safety regulations to eliminate risks.

Table 3. Incidents per 10,000 miles (16,093 km) of pipeline, 2000–2019.

Year	Incidents per 10,000 miles
2000	2.21
2001	2.36
2002	1.92
2003	2.74
2004	2.80
2005	3.60
2006	3.68
2007	2.92
2008	3.13
2009	3.08
2010	2.74
2011	3.37
2012	2.95
2013	3.15
2014	3.93
2015	4.27
2016	2.79
2017	3.16
2018	3.59
2019	3.85 [a]

[a] The PHMSA figure for miles of onshore transmission pipelines for natural gas in 2019 is not yet available, so this figure was estimated using the miles of pipelines in 2018.

Table 4 provides the results of the primary regression. Except for the *Midwest* and *South* dummy variables, the effects of these variables are statistically significant at the 95% confidence level. Population density and the natural logs of mean income have negative and significant coefficients, showing that the influence of these variables on safety precautions dominates the influence of population density and income on the proximity of people and buildings to the pipeline, as discussed in Section III. This is the case holding constant the real estate taxes per square mile, a gauge for the value of property in the area. When the real estate tax variable is removed, as shown in Table 5, the significance of mean income falls, perhaps because mean income becomes a proxy for both more inspections (a negative influence) and more valuable property (a positive influence).

Table 4. Regression results (dependent variable: Cost of incident).

Independent Variable	Coefficient	t-Ratio
Population density	−0.0001	−6.98
Log mean income	−0.544	−2.60
East	−0.564	−2.36
Midwest	−0.022	−0.11
South	−0.185	−1.03
Log pipeline age	0.222	4.32
Log real estate taxes	0.057	2.71
Constant	16.824	7.51

Table 5. Additional regressions (dependent variable: Cost of incident. t-values are in parentheses).

Independent Variable	Regression				
	1 log-log	2 linear	3 log-log with % farms	4 log-log no taxes	5 log-log prop. cost only
Population density	−0.0001 (−6.98)	−325.829 (−1.73)	−0.0001 (−6.71)	−0.0001 (−6.48)	−0.0001 (−6.55)
Mean income	−0.544 (−2.60)	−51.66 (−1.78)	−0.319 (−1.70)	−0.291 (−1.55)	−0.504 (−2.27)
East	−0.564 (−2.36)	−6600347 (−2.33)	−0.559 (−2.33)	−0.527 (−2.20)	−0.364 (−1.44)
Midwest	−0.022 (−0.11)	−5090143 (−2.22)	0.071 (0.35)	−0.034 (−0.17)	0.123 (0.60)
South	−0.185 (−1.03)	−4858490 (−2.29)	−0.141 (−0.77)	−0.218 (−1.21)	−0.148 (−0.78)
Pipeline age	0.222 (4.32)	11908.14 (0.40)	0.218 (4.24)	0.220 (4.27)	0.264 (4.85)
Real estate taxes per mile	0.057 (2.71)	4.017432 (5.53)			0.054 (2.45)
Percent farms			−0.129 (−2.08)		
Constant	16.824 (7.51)	7692046 (2.67)	15.018 (7.17)	14.587 (6.98)	16.07917 (6.79)
R-squared	0.0743	0.0356	0.0718	0.0681	0.0700

The negative coefficients on *East*, *Midwest*, and *South* were expected given the relatively large cost of incidents in the West, as apparent from Table 2. As hypothesized, the coefficient on the *Log real estate taxes* variable was positive and significant at the 95% level, showing that in areas with relatively valuable real estate, and thus larger yields for real estate taxes, an incident causes more costly damage.

The coefficient on *Log pipeline age* indicates that a one percent increase in the age of a pipeline corresponds to a 0.222% increase in the expected cost of a pipeline incident. Applying that to the average cost of a pipeline incident, a one percent increase in pipeline age represents an increase of $3293 in the cost of the average incident along that pipeline.

The influence of pipeline age on damage costs is relevant to potential host communities for several reasons. To the extent that newer pipelines are safer than older pipelines, new projects have lower expected damage costs than existing projects. The rate of decline in pipeline safety over time is also relevant to communities as they consider the prospect of incidents well into the future, when the character of the community and its level of development may change. In addition, the risks associated with old pipelines must be considered when new projects involve the repurposing of existing pipelines. The average year of installation for a pipeline involved in an incident since 2000 is 1973.

Overall, the regression analysis reveals the indiscriminate nature of damages from pipeline incidents. The R-squared indicates that 7.4% of the variation in costs is caused by the variables in the equation. So even factoring in the influence of these variables, there is considerable uncertainty about the cost imposed by a leak or explosion. The largest sources of variation are specific to individual cases and are not captured by the variables in the dataset. This motivates communities' need for additional forms of insurance to mitigate risk and uncertainty, as discussed in Section 4.

Several versions of the regression equation were estimated to test the robustness of the findings. Table 5 shows the results. The signs on the coefficients and their significance are largely consistent with a few exceptions. Regression 1 repeats the findings discussed above for the purpose of comparison. Regression 2 is a linear version of the specification, which is an inferior fit but demonstrates the robustness of the findings. Regression 3 substitutes the percent of farms for the real estate taxes per square mile. Like the real estate taxes variable, the percent farms variable has a negative coefficient and is significant at the 95% level. Using both of those variables lowers the adjusted R-squared and causes both variables to lose their significance at the 95% level. Regression 4 includes neither real estate taxes nor percent farms, yielding results similar to the other regressions but an inferior fit. Regression 5 provides coefficients for the specification with a dependent variable of property cost only, rather than total cost.

5. Discussion

The results indicate that incidents along onshore natural gas transmission lines represent a small fraction—15%—of all pipeline incidents reported to the PHMSA between 2000 and 2019. Over the past decade, an average of 101 such incidents occurred annually in the United States. When an incident occurs, the damage can devastate the local community. Compensation for lost lives, bodily injuries, property damage, environmental damage, and related expenses are often subject to litigation. Pipelines can also create fears and anxieties in communities that go uncompensated. The findings of this research give communities a better idea of the scale and frequency of relevant incidents and quantify identifiable influences on damage costs.

Current approaches to pipeline safety focus on regulation. For example, in response to deadly incidents along onshore gas pipelines, the PHMSA tightened its integrity management requirements in 2019. The new rules require pipeline operators to take further precautions, such as additional monitoring of the pressure in natural gas transmission pipelines and more assessments of pipelines in areas that are populated but not designated as high-consequence areas [22]. Such regulations are valuable attempts to increase pipeline safety, but they do not assist the victims of pipeline accidents when, despite the regulations, they occur.

To serve both pipeline companies and host communities well, policymakers must attend to the dual realities of low-incident probabilities and high costs for the rare victims. Solutions should also address the troubling uncertainty for pipeline hosts. The five worst onshore natural gas transmission pipeline incidents over the past decade each caused more than $25 million worth of property damage [4]. All damage is disruptive to the property owners and victims, whether compensation is provided or not. Incidents involving explosions generate inordinate media attention and corresponding fears and concerns.

Communities would benefit from the certainty of insurance against the downside risk of a pipeline leak or explosion. One solution is for pipeline operators to act as insurers. These firms have a relatively clear understanding of the risks. They also make decisions that influence the pipelines' safety, meaning there are beneficial incentive effects of pipeline operators serving as insurers. Operators could budget for the expected cost with knowledge of the pipeline's history, the safety measures in place, and the monitoring practices, and they provide certain compensation when problems occur.

The pipeline operators are able to spread the risk of a costly incident across their entire pipeline network. In his concurring opinion on the legal case of *Escola* vs. *Coca Cola* [23], Justice Roger J. Traynor remarked on the ability of companies such as Coca Cola to spread the risk of injuries caused by their products broadly as a cost of doing business. Already, many of the costs of pipeline incidents are covered by the pipeline operators as payments to communities and damage awards in litigation. If full compensation of host communities became mandated or contractual, the pipeline owners would provide certainty where it is needed. In addition, by internalizing the external costs of their decisions, baring other sources of market failure, firms would make socially optimal decisions about pipeline construction and use.

Personal injuries have explicit and implicit values and require special consideration. In the case of lost human life, no amount of ex-post compensation is enough. However, we can apply the value of a statistical life—the life of an unidentified individual whose death we can anticipate due to the risks inherent in pipeline use. The literature review explains that the estimated value of a statistical injury is $90,697 and the estimated value of a statistical life is $10.45 million. If guaranteed in advance, compensation at these levels would provide appropriate incentives for firms and fitting ex-ante assurance for communities.

One form of the insurance remedy would be an application of the precautionary polluter pays principle [24]. The pipeline owners could create a trust fund with the amount that would compensate the victims for the worst-case scenario. That amount would go to the community in the event of damage and the trust fund would be replenished. If no damage occurred over the lifetime of the pipeline, the money in the fund would be returned to the pipeline owner. The results of this study regarding the expected cost and the influence of pipeline age, among other variables, would be informative for any such solution.

6. Conclusions

Expansive pipeline networks carry natural gas from source to use. Communities grapple with their stance on these conduits and need specific information to do so prudently. This article examines the costs communities face in the particular case of hosting natural gas transmission lines in the United States. Several community characteristics have a statistically significant effect on the cost of pipeline incidents, as does the age of the pipeline.

Reportable incidents along onshore natural gas transmission pipelines occur about three times per 10,000 miles (16,093 km) of pipeline per year. The low probability of an incident is coupled with the potential for catastrophic harm. Extensive media coverage of the worst disasters exacerbates community fears. The resulting uncertainty leaves many communities discomforted by the prospect of hosting a pipeline. Without remedies for the uncertainty, both full information and safety regulations fall short of solving the problem.

Given the ongoing pattern of tragic pipeline incidents, communities need solutions that provide certain compensation. Possibilities include variants on the precautionary polluter pays principle. This would place the burden on those most informed, most able to minimize the risks, and most able to spread the risks broadly across many communities. This approach would alleviate uncertainty for communities and remove the need for costly litigation over compensation for damages. This type of solution might also reduce the need for some other regulatory measures because it causes the pipeline operators to internalize the external costs of risky behavior. The remedies discussed here could apply similarly to pipelines carrying other substances in any country.

Funding: This research received no external funding.

Acknowledgments: The author thanks Gabrielle Gilkison and Skyler Palmer for excellent research assistance.

Conflicts of Interest: The author declares no conflict of interest.

References

1. Van Velzer, R. Controversial Kentucky Pipeline Conversion Project Scrapped. Available online: https://wfpl.org/controversial-kentucky-pipeline-conversion-project-scrapped/ (accessed on 23 March 2020).
2. The Inter-Mountain. Community Welcomes Pipeline Employees with Special Event. Available online: https://www.theintermountain.com/news/communities/2018/07/community-welcomes-pipeline-employees-with-special-event/ (accessed on 17 March 2020).
3. Energy Information Administration. Natural Gas. Available online: https://www.eia.gov/dnav/ng/ng_cons_sum_dcu_nus_a.htm (accessed on 23 March 2020).
4. PHMSA. Pipeline Incident 20 Year Trends. Available online: https://www.phmsa.dot.gov/data-and-statistics/pipeline/pipeline-incident-20-year-trends (accessed on 23 March 2020).
5. Wilde, L.; Loos, C.; Williamson, J. Pipelines and property values: An eclectic review of the literature. *J. Real Estate Lit.* **2012**, *20*, 245–260.
6. McElveen, M.A.; Brown, B.E.; Gibbons, C.M. Natural gas pipelines and the value of nearby homes: A spatial analysis. *J. Hous. Res.* **2017**, *26*, 27–38.
7. Integra Realty Resources. *Pipeline Impact to Property Value and Property Insurability*; Report No. 2016.01; INGAA Foundation: Washington, DC, USA, 2016.
8. Simons, R.A.; Winson-Geideman, K.; Mikelbank, B.A. The effects of an oil pipeline rupture on single-family house prices. *Apprais. J.* **2001**, *69*, 410–418.

9. Hansen, J.L.; Benson, D.; Hagen, A. Environmental hazards and residential property values: Evidence from a major pipeline event. *Land Econ.* **2006**, *82*, 529–541. [CrossRef]
10. Kielisch, K. *Study on the Impact of Natural Gas Transmission Pipelines*; Forensic Appraisal Group, Ltd.: Neenah, WI, USA, 2015.
11. Herrnstadt, E.; Sweeney, R.L. *What Lies Beneath: Pipeline Awareness and Aversion*; NBER Work. Papers 23858; NBER: Cambridge, MA, USA, 2017. [CrossRef]
12. Fang, W.; Wu, J.; Bai, Y.; Zhang, L.; Reniers, G. Quantitative risk assessment of a natural gas pipeline in an underground utility tunnel. *Process Saf. Prog.* **2019**, *38*, e12051. [CrossRef]
13. Jo, Y.D.; Park, K.S.; Kim, H.S.; Kim, J.J.; Kim, J.Y.; Ko, J.W. A quantitative risk analysis method for the natural gas pipeline network. *Trans. Inf. Commun. Technol.* **2010**, *43*, 195–203. [CrossRef]
14. Han, Z.Y.; Weng, W.G. An integrated quantitative risk analysis method for natural gas pipeline network. *J. Loss Preve. Process Ind.* **2010**, *23*, 428–436. [CrossRef]
15. Viscusi, W.K. The Role of Publication Selection Bias in Estimates of the Value of a Statistical Life. *Am. J. Health Econ.* **2015**, *1*, 27–52. [CrossRef]
16. Viscusi, W.K.; Aldy, J.E. The Value of a Statistical Life: A Critical Review of Market Estimates throughout the World. *J. Risk Uncertain.* **2003**, *27*, 5–76. [CrossRef]
17. PHMSA. Data and Statistics Overview. Available online: https://www.phmsa.dot.gov/data-and-statistics/pipeline/data-and-statistics-overview (accessed on 1 March 2020).
18. PHMSA. Pipeline Facility Incident Report Criteria History. Available online: https://www.phmsa.dot.gov/data-and-statistics/pipeline/pipeline-facility-incident-report-criteria-history (accessed on 9 March 2020).
19. Restrepo, C.E.; Simonoff, J.S.; Zimmerman, R. Causes, cost consequences, and risk applications of accidents in US hazardous liquid pipeline infrastructure. *Int. J. Crit. Infrastruct. Prot.* **2009**, *2*, 38–50. [CrossRef]
20. PHMSA. *Pipeline Safety: Class Location Requirements*; 78 FR 46560; PHMSA: Washington, DC, USA, 2013.
21. Pless, J. Making State Gas Pipelines Safe and Reliable: An Assessment of State Policy. In Proceedings of the National Conference of State Legislatures, March 2011. Available online: https://www.ncsl.org/research/energy/state-gas-pipelines-natural-gas-as-an-expanding.aspx#Population_Density___Pipeline_Mileage_Per_Square_Foot_of_Land (accessed on 22 March 2020).
22. PHMSA. Pipeline Safety: Safety of Gas Transmission Pipelines: MAOP Reconfirmation, Expansion of Assessment Requirements, and Other Related Amendments; 84 FR 52180; 2019. Available online: https://www.federalregister.gov/documents/2019/10/01/2019-20306/pipeline-safety-safety-of-gas-transmission-pipelines-maop-reconfirmation-expansion-of-assessment (accessed on 11 April 2020).
23. Traynor, R.J. Escola v. Coca Cola Bottling Co. 1944. Available online: https://repository.uchastings.edu/traynor_opinions/151/ (accessed on 11 April 2020).
24. Anderson, D. *Environmental Economics and Natural, Resource Management*, 5th ed.; Routledge: New York, NY, USA, 2019. [CrossRef]

© 2020 by the author. Licensee MDPI, Basel, Switzerland. This article is an open access article distributed under the terms and conditions of the Creative Commons Attribution (CC BY) license (http://creativecommons.org/licenses/by/4.0/).

Article

The Response of US Macroeconomic Aggregates to Price Shocks in Crude Oil vs. Natural Gas

Jin Shang and Shigeyuki Hamori *

Graduate School of Economics, Kobe University 2-1, Rokkodai, Nada-Ku, Kobe 657-8501, Japan; susanfeir@gmail.com
* Correspondence: hamori@econ.kobe-u.ac.jp; Tel.: +81-080-9309-6868

Received: 29 March 2020; Accepted: 15 May 2020; Published: 20 May 2020

Abstract: Price fluctuations in crude oil and natural gas, as important sources of energy, have a remarkable influence on our economies and daily lives. Therefore, it is extremely important to react appropriately and to formulate appropriate policies or strategies to reduce the expected negative effects of fluctuations. However, as Kilian suggested, not all oil price shocks are similar; price increases can have diverse impacts on the real price of oil, depending on the underlying determinants of the price fluctuation. Therefore, economists, policymakers, and investors need to decompose real price shocks and evaluate the responses of macroeconomic aggregates to different types of shocks. In this study, we investigate and compare the different effects crude oil and natural gas price shocks have on US real GDP and CPI levels, utilizing a two-stage method based on a structural vector autoregression (SVAR) model proposed by Kilian. We found that a crude oil specific demand shock made larger contributions to the real price of oil than a natural gas specific demand shock did to the real price of gas, and that specific demand shocks in crude oil and natural gas markets had different effects on US CPI inflation and had similar effects on the real US GDP level.

Keywords: SVAR; oil price; gas price; US macroeconomic aggregates; GDP; CPI

1. Introduction

Researchers have studied crude oil for many years because it is of major interest as a significant but limited resource. In addition, the crude oil index plays an important role among economists, policymakers, and investors, because fluctuations in either its price or its production have a remarkable impact on the world economy and stock markets. On the contrary, as a relatively environmentally friendly resource and a key fuel for the electrical power and industry sectors, natural gas has attracted increasing attention. Evidence for this increasing attraction comes from the fact that the consumption of natural gas has risen to a share of 23% of total energy resources, and is the fastest-growing fossil fuel among energy resources [1]. Another fact is that the demand for natural gas increased by 4.6% in 2018, accounting for almost half of all energy resource demand growth [2]. Thus, it is not difficult to conclude that fluctuations in the prices of crude oil and natural gas greatly influence economic activity and daily life.

Researches had shown that there is a correlation between US economic growth and exogenous oil shocks. For instance, Hamilton [3,4] and Hooker [5] proposed that after oil price increases, there were recessions in the US economy. However, as proposed by Kilian [6], depending on the underlying causes, such as supply shocks, aggregate demand, or precautionary demand, price fluctuations have varying dynamic effects on real prices, and hence on the economy. Other studies have agreed; for instance, Kling [7] reported that the association is obvious between crude oil price increases and declines in the stock market. However, Kilian and Park [8] reported that US real stock returns reacted differently to oil price shocks, depending on the origins driving the oil price shocks. Other research has also analyzed the macroeconomic effects of oil market fluctuations. Oladosu et al. [9] adopted a quantitative meta-regression model to simulate the oil price elasticity of GDP in US and their results revealed that the estimated US GDP elasticity was negative, and particularly smaller than about 10 years ago. Jan van de Ven and Fouquet [10] identified the impact of energy shocks on economic activity using data from the United Kingdom for 310 years and their study indicated that the influences of supply shocks declined due to the partial shift from coal to crude oil, and that it is possible to reduce vulnerability and increase resilience by substituting renewable energy sources for traditional energy sources. Ju et al. [11] used data from 1980 to 2014 to investigate the macroeconomic performance of oil price shocks by adopting three methods, namely, empirical covariance, robust covariance, and support vector machine methods, and the results implied that the outlier performances of GDP, CPI, and the unemployment rate are consistent with the oil price shock process. Correspondingly, some literature investigated the relationship between natural gas price shocks and macroeconomic aggregates. Nick and Thoenes [12] extended the structural vector autoregression (SVAR) model and analyzed the effects of different fundamental influences on the price of natural gas in the German market. Their results revealed that in the short-term, the price of natural gas tends to be affected by some factors, such as temperature, inventory, and supply insufficiencies, and, by contrast, in the long-term the price tends to be influenced by the overall economic climate and the substitutional relationship between crude oil and coal. Zhang et al. [13] used a computable general equilibrium model to investigate the macroeconomic effects of natural gas prices in China and their results showed that when natural gas prices increase, the CPI would increase and GDP would decrease. In addition, research has also emphasized the transmission effect of oil demand and supply shocks on the natural gas market. For example, Jadidzadeh and Serletis [14] analyzed the effects of supply-demand shocks stemming from the global oil market on the real price of natural gas and found that oil supply-demand shocks accounted for approximately 45% of the fluctuation in the price of natural gas.

This study's first objective is to decompose shocks to the real price of crude oil and natural gas into three components: (1) oil/gas supply shocks, (2) global aggregate demand shocks, and (3) specific or precautionary demand shocks. Its second objective is to evaluate important differences in how US macroeconomic aggregates, such as CPI and real GDP growth, react to various oil shocks underlying the real price of crude oil and natural gas. The third objective of this study is to compare its results with Kilian's results regarding crude oil in 2009, in order to determine whether conditions have changed since 2007. Our data cover the period from 1973:1 to 2019:6, whereas Kilian [6] only had data up to 2007:12. Its fourth objective is to compare the impacts on crude oil and natural gas of different types of shocks to US macroeconomic aggregates. We expect this study to have implications for market operators and investors.

Because one of the objectives of our study is to compare our results with Kilian's work, and because we used the same approach as Kilian [6], there is a necessity to demonstrate the similarities and discrepancies between the two works. We used the same methodology and variables as in Kilian's work. However, because the importance and the potentiality of natural gas has gradually increased, we added natural gas to our study to check the effects of each structural shock on the real price of oil and gas and compare their effects on the US macroeconomic aggregates; in contrast, Kilian's work focused on the crude oil market. Another difference is that our sample period of crude oil is longer than Kilian's to verify if there are any changes after 2007.

Our findings in this study are threefold. First, by constructing an SVAR model to quantify the responses to one-standard-deviation structural shocks, we found that there were differences between the effects on crude oil and natural gas in the response patterns of real economic activity due to precautionary demand shocks. Second, by decomposing historic real prices to check the cumulative contribution from each demand and supply shock to the real price, we found that the cumulative effect of precautionary demand shocks was varying in degree between crude oil and natural gas; specifically, the cumulative contribution of an oil-specific demand shock to the real oil prices is larger than the cumulative contribution of a gas-specific demand shock to the real gas prices. Third, by utilizing the regression model to investigate how oil and gas demand and supply shocks that underlie the real price of oil and gas influence US macroeconomic aggregates, we found that precautionary demand shocks on crude oil and natural gas have different effects on CPI inflation and had similar effects on real GDP growth. Both oil- and gas-specific demand shocks led to a small, but statistically insignificant, reduction of the US GDP level. Oil-specific demand shocks tended to cause a large and statistically significant increase in CPI inflation; by contrast, gas market-specific demand shocks led to a small and statistically insignificant increase in the US CPI level. This result implied policymakers should react differently to gas-specific demand shocks and oil-specific demand shocks.

The remainder of this paper is organized as follows. Section 2 provides a detailed description of the data and introduces the econometric model, a two-stage method based on the SVAR model proposed by Kilian [6]. Section 3 identifies the structural shocks that drive the real price of crude oil and natural gas. We quantify the historical evolution of these shocks and the response to these shocks from production, real activity, and the real price of crude oil and natural gas. We decompose the real price of crude oil and natural gas over time to assess the respective cumulative contribution of each shock to real prices. Finally, we analyze the effects of those shocks on US macroeconomic aggregates. Section 4 concludes the study.

2. Materials and Methods

2.1. Data

Table 1 describes the datasets we utilized and their sources. These datasets included crude oil production, gas production, a real economic activity index (the Kilian index available on his website), crude oil prices, natural gas prices, and US real GDP, deflated using US CPI (Kilian [6]).

Table 1. Data descriptions and sources.

Variable	Sample Period	Frequency	Description	Data Source
$\Delta oprod_t$	1973:1–2019:6	monthly	Percentage change in global crude oil production	IEA [2]
l_rea_t	1973:1–2019:6	monthly	Logarithmic real economic activity index (Kilian index)	Lutz Kilian website [15]
l_rpo_t	1973:1–2019:6	monthly	Log real price of crude oil [1]	Federal Reserve Bank
$\Delta gprod_t$	1994:1–2019:6	monthly	Percentage change in global natural gas production	IEA [2]
l_rpg_t	1994:1–2019:6	monthly	Log real price of natural gas [1]	Federal Reserve Bank
gdp_gr	1973:1–2019:6	quarterly	Real US GDP growth	OECD
cpi	1973:1–2019:6	quarterly	US CPI	OECD

[1] Real price means price which is deflated by US CPI. [2] Data from the International Energy Agency.

We collected data from Bloomberg, Thomson Reuters, and OECD Stat. These datasets include the crude oil spot price (West Texas Intermediate, WTI) and the natural gas spot price (Henry Hub). l_rpo_t defers to the real price of crude oil expressed in log terms. l_rpg_t defers to the real price of natural gas expressed in log terms. Regarding the dataset of the Kilian index, we collected the data from the Lutz Kilian website. From Kilian (2009), the Kilian index is a detrended real freight rate index constructed as a measure of monthly global real economic activity, which could imply the variation of worldwide real economic activity and capture the shifts in the demand for industrial commodities throughout international markets [6,15] (see Appendix A for details). In this study, l_rea_t defers to the Kilian index; because the minimum of the Kilian index is −161.643, we added 196 to every data point and expressed the series in log terms. We collected production data for oil and gas from the International Energy Agency (IEA), and calculated the change rate of oil/gas production in percentage terms to reflect the percentage change of worldwide oil/gas production. $\Delta oprod_t$ refers to the percentage change in global crude oil production, and $\Delta gprod_t$ refers to the percentage change in global natural gas production. US real GDP growth and US CPI data was collected from the Organization for Economic Co-operation and Development (OECD); because data for the US real GDP growth rate is unavailable at a monthly frequency, we adopted the quarterly frequency when we downloaded the US real GDP growth rate and US CPI datasets. The sample period for crude oil data was from January 1973 to June 2019, but the natural gas sample was limited by the available data period of January 1994 to June 2019.

The descriptive statistics and stationarity tests for the variables are shown in Table 2. This table indicates the mean, maximum, minimum, and standard deviation of each variable. (Although the log of the real price of natural gas may have a unit root, we checked the stability condition of the SVAR system and found that the system is stationary; more details are provided in Appendix B.)

Table 2. The results of unit root tests for the variables.

	$\Delta oprod_t$	l_rea_t	l_rpo_t	$\Delta gprod_t$	l_rpg_t	gdp_gr	cpi
Mean	0.00086	5.24667	3.34438	0.00270	1.25432	0.67169	153.274
Maximum	0.06714	5.95382	4.84602	0.05266	2.56832	3.86425	255.310
Minimum	−0.09432	3.53681	1.20789	−0.04827	0.32930	−2.16381	42.7000
Std. Dev.	0.01502	0.28056	0.72775	0.01415	0.48039	0.77499	61.5513
Unit root test t-stat.	−25.426	−4.356	−2.893	−15.103	−2.473	−9.62668	−10.238
Unit root test Prob.	0.000	0.000	0.047	0.000	0.123	0.000	0.000

2.2. Methodology

Based on the two-stage method proposed by Kilian [6], first we identified the structural shocks by estimating the SVAR model and decomposing the structural shocks. Second, we constructed a regression model to investigate how the structural shocks affected US macroeconomic aggregates such as real GDP growth and CPI inflation. This methodology is widely used in investigations related to oil and natural gas price shocks. For instance, Sim and Zhou [16] applied a structural VAR model and a quantile regression model to examine the relationship between oil prices and US stock returns. Yoshizaki and Hamori [17] utilized the same model to analyze the reaction of economic activity, inflation, and exchange rates to oil prices shocks. Ahmed and Wadud [18] also used a structural VAR model to identify the role of oil price shocks on the Malaysian economy and monetary responses.

2.2.1. The SVAR Model

Consider an SVAR model based on a monthly data set for crude oil, $o_t = (\Delta oprod_t, rea_t, rpo_t)$, where $\Delta oprod_t$ represents the percentage change in global crude oil production, rea_t represents an index of real economic activity, and rpo_t represents the real price of crude oil. In addition, consider an SVAR model for natural gas, $g_t = (\Delta gprod_t, rea_t, rpg_t)$, where $\Delta gprod_t$ represents the percentage change in global natural gas production, rea_t represents an index of real economic activity, and rpg_t represents the real price of natural gas.

Thus, the representation of the SVAR model of crude oil is:

$$A_0 o_t = \alpha + \sum_{i}^{24} A_i o_{t-i} + \varepsilon_t \qquad (1)$$

where ε_t represents the vector of mutually and serially uncorrelated structural innovations. Although the lag length indicated by Akaike's information criterion was 14, we decided to use 24, as also used by Kilian, because we used monthly data series in the model. Using 24 lags avoids the dynamic misspecification problem [6]. When the error terms are related, they have a common component that cannot be recognized by any particular variable, thus, we performed an adjustment to make the error terms orthogonal by structural decomposition. We assumed that A_0^{-1} has a recursive structure; therefore, we can decompose the errors e_t according to $e_t = A_0^{-1}\varepsilon_t$:

$$e_t = \begin{pmatrix} e_t^{\Delta oprod} \\ e_t^{rea} \\ e_t^{rpo} \end{pmatrix} = \begin{bmatrix} a_{11} & 0 & 0 \\ a_{21} & a_{22} & 0 \\ a_{31} & a_{32} & a_{33} \end{bmatrix} \begin{pmatrix} \varepsilon_t^{oil\ supply\ shock} \\ \varepsilon_t^{aggreagate\ demand\ shock} \\ \varepsilon_t^{oil-specific\ demand\ shock} \end{pmatrix} \qquad (2)$$

Based on Equation (2), we have three structural shocks to be identified, namely an oil supply shock, an aggregate demand shock, and an oil-specific demand shock. As noted by Kilian [6], an oil supply shock is designed to capture unexpected innovations to international oil output. An aggregate demand shock, which is driven by the global business cycle, is designed to identify the unexpected innovations to global economic activity. Finally, an oil-specific demand shock, or precautionary demand shock, which is driven by increasing uncertainty in the oil market, is designed to identify the exogenous shifts in precautionary demand.

It must be noted that, as Equation (2) shows, in the first row of A_0^{-1}, the restriction $a_{12} = a_{13} = 0$ implies that innovations to global oil production can only be explained by the oil supply shocks; in the middle row, the restriction $a_{23} = 0$ indicates that innovations to worldwide economic activity can be explained by oil supply and aggregate demand shocks rather than an oil-specific demand shock. The restriction of the last row shows that all of the three shocks could have effects on the real price of crude oil.

Therefore, we can obtain the structural residuals by estimating from Equations (1) and (2). Here we can plot the structural residuals to observe the changing composition of the structural shocks over time. Then, we can check the dynamic response pattern of the endogenous variables to various structural shocks by imposing a one-standard-deviation structural shock.

Similarly, the representation of the SVAR model of natural gas is:

$$A_0 g_t = \beta + \sum_{i}^{24} A_i g_{t-i} + w_t \qquad (3)$$

We can also decompose the errors w_t according to $w_t = A_0^{-1} \omega_t$ to identify the structural shocks:

$$w_t = \begin{pmatrix} \omega_t^{\Delta oprod} \\ \omega_t^{rea} \\ \omega_t^{rpo} \end{pmatrix} = \begin{bmatrix} a_{11} & 0 & 0 \\ a_{21} & a_{22} & 0 \\ a_{31} & a_{32} & a_{33} \end{bmatrix} \begin{pmatrix} \omega_t^{gas\ supply\ shock} \\ \omega_t^{aggreagate\ demand\ shock} \\ \omega_t^{gas-specific\ demand\ shock} \end{pmatrix} \qquad (4)$$

2.2.2. Regression Model

The purpose of this step is to investigate how the structural shocks estimated from the structural VAR model in Section 2.2.1 influence the US macroeconomic aggregates, such as CPI inflation (π_t) or real GDP growth (Δy_t). The effects are identified by the following regressions:

$$\Delta y_t = \alpha_j + \sum_{i=0}^{12} \phi_{ji} \hat{\zeta}_{jt-i} + u_{jt} \qquad (5)$$

$$\pi_t = \delta_j + \sum_{i=0}^{12} \psi_{ji} \hat{\zeta}_{jt-i} + v_{jt} \qquad (6)$$

$$\hat{\zeta}_{jt}^o = \frac{1}{3} \sum_{i=1}^{3} \hat{\varepsilon}_{j,t,i}, \ j = 1, 2, 3 \qquad (7)$$

$$\zeta_{jt}^g = \frac{1}{3}\sum_{i=1}^{3} \hat{\omega}_{j,t,i}, \ j=1,2,3 \tag{8}$$

In this regression model, u_{jt} and v_{jt} refer to errors; ϕ_{ji} and ψ_{ji} represent the impulse response coefficients, respectively; $\hat{\varepsilon}_{j,t,i}$ denotes the estimated residual for the j_{th} structural shock in the i_{th} month of the t_{th} quarter in the crude oil data; and $\hat{\omega}_{j,t,i}$ denotes the estimated residual for the j_{th} structural shock in the i_{th} month of the t_{th} quarter in the natural gas data. Furthermore, we determined the number of lags to be 12 quarters as also used by Kilian [6].

3. Results

3.1. Identifying Structural Shocks

Figures 1–4 summarize the empirical results of the SVAR model based on Equations (1)–(4) from Section 2. After obtaining the model estimates, we can calculate the structural residuals.

3.1.1. Quantifying the Evolution of Crude Oil and Natural Gas Demand and Supply Shocks

Figure 1 plots the time path of the structural shocks estimated by the model.

Figure 1 reveals that, at any point in time, the real price of crude oil was reacting to multiple shocks, the combination of which evolved:

- In 2008, the oil side results were characterized by a sharp plummet due to the global aggregate demand shock associated with the 2008 financial crisis. In 2008, the oil-specific structural demand shock also led to a sharp fall associated with the dramatic increase in the price of crude oil. Accordingly, the demand for natural gas increased, leading to a substantial and positive gas-specific structural demand shock.
- One point to note is that because the time periods for the crude oil and natural gas structural shocks did not overlap, it is inappropriate to conclude that the amplitude and frequency of crude oil structural shocks were greater than those of natural gas structural shocks.

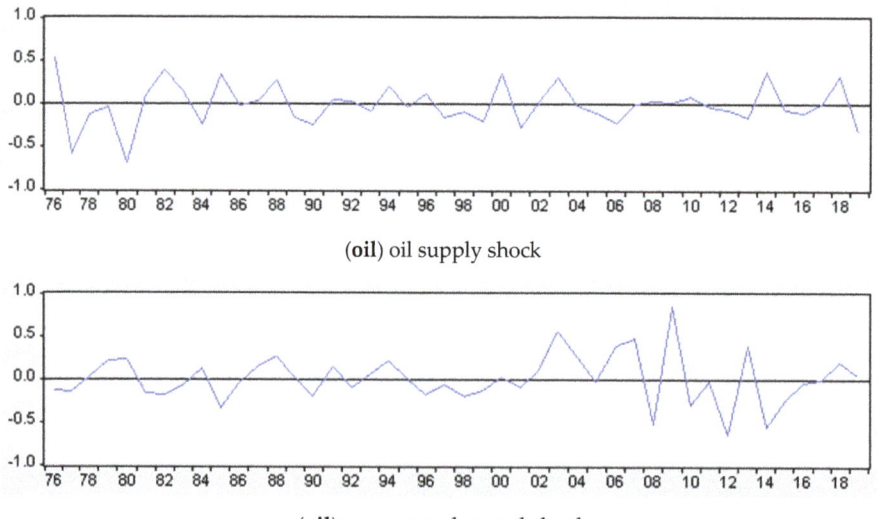

Figure 1. Cont.

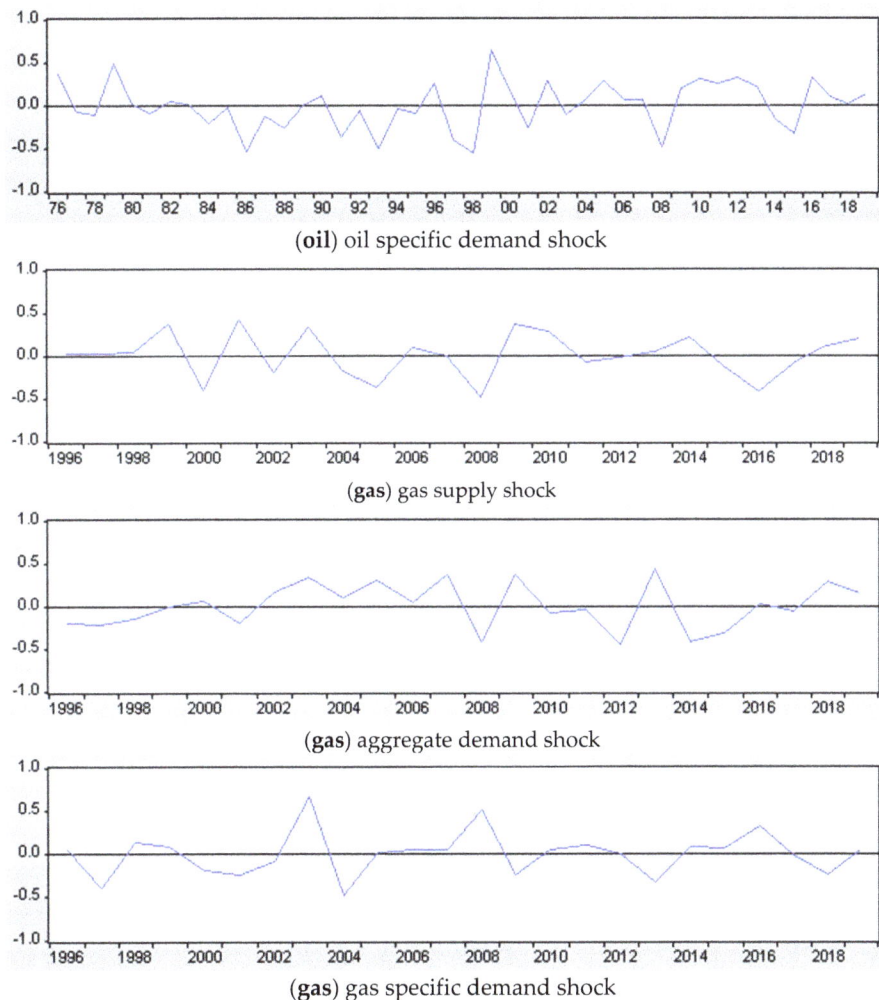

Figure 1. Historical evolution of structural shocks. From top to bottom, column (**oil**) illustrates crude oil supply, aggregate demand, and oil-specific demand shocks from 1976 to 2019. From top to bottom, column (**gas**) illustrates natural gas supply, aggregate demand, and gas-specific demand shocks from 1996 to 2019. **Note:** Structural residuals implied by Equations (1)–(4), averaged annually.

3.1.2. Responses to One-Standard-Deviation Structural Shocks

Figure 2 plots the responses of global oil and gas production, real economic activity, and the real price of oil and gas to a one-standard-deviation structural innovation.

Figure 2 reveals how global oil and gas production, real economic activity, and the real price of oil and gas responded to demand and supply shocks, respectively, in the crude oil and natural gas markets.

First, the empirical results for crude oil are as follows: Concerning oil supply shocks, we can interpret from the graph that positive oil supply shocks, in which production rose, caused a sharp increase, followed by a reversal of the increase to its initial level within the first quarter. At the same time, one-standard-deviation oil supply shocks on the real price of oil and real economic activity are statistically insignificant at all observed time horizons observed.

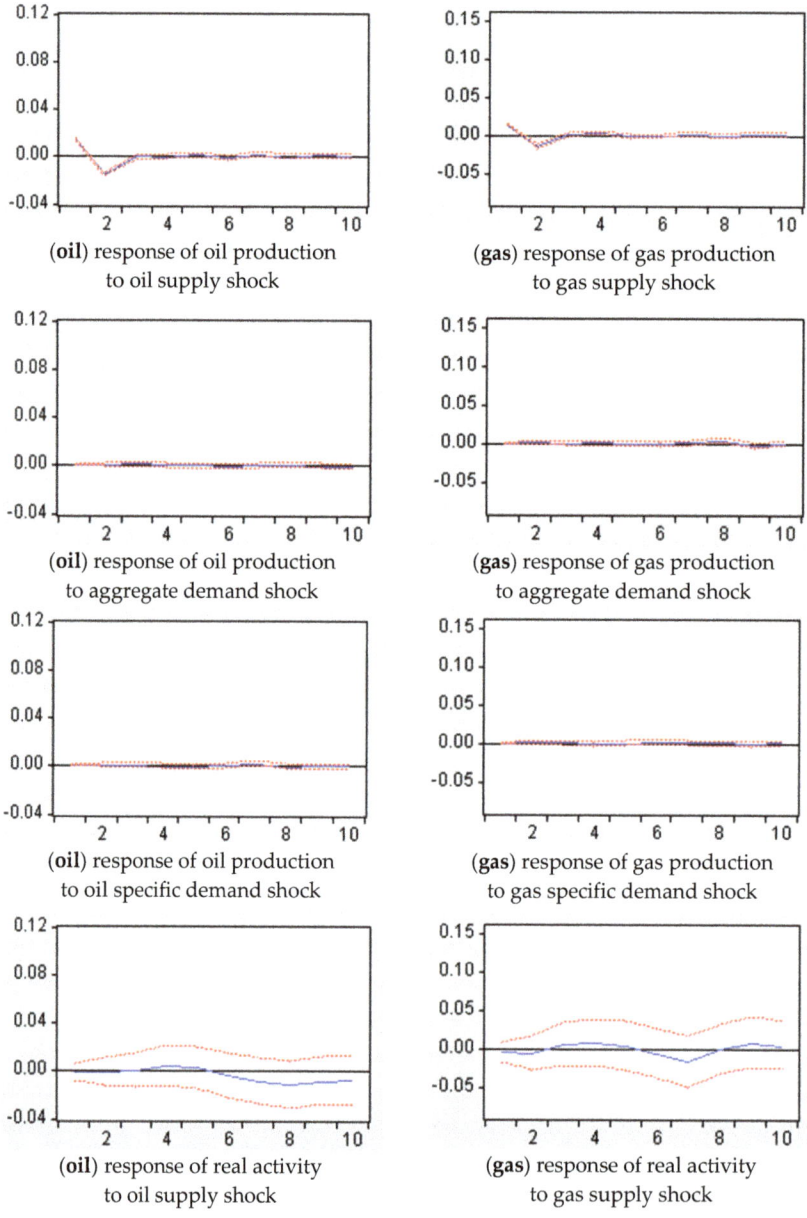

Figure 2. Cont.

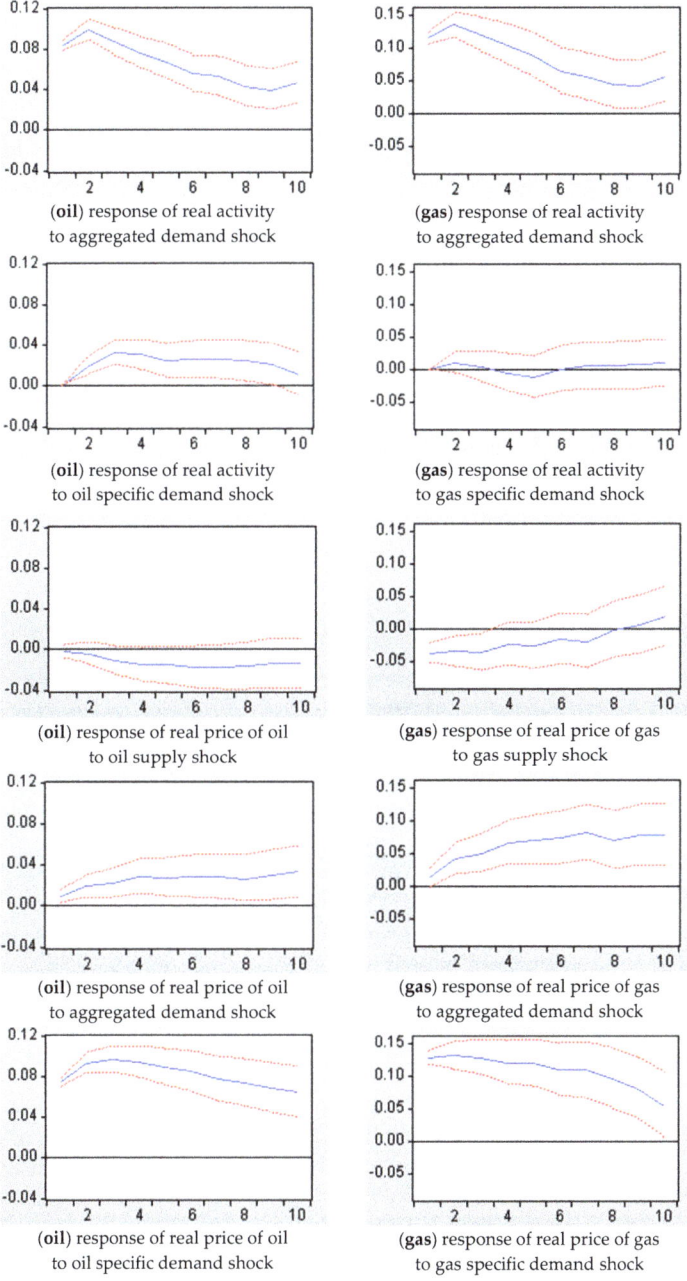

Figure 2. Responses to one-standard-deviation structural shocks. From top to bottom, the first column in (**oil**) illustrates the responses of crude oil production, real economic activity, and the real price of oil, to oil supply shocks, aggregate demand shock, and oil-specific demand shock, respectively. From top to bottom, the second column in (**gas**) illustrates the responses of natural gas production, real economic activity, and real price of gas to gas supply shocks, aggregate demand shock, and gas-specific demand shock, respectively. **Note:** Estimates based on Equations (1)–(4).

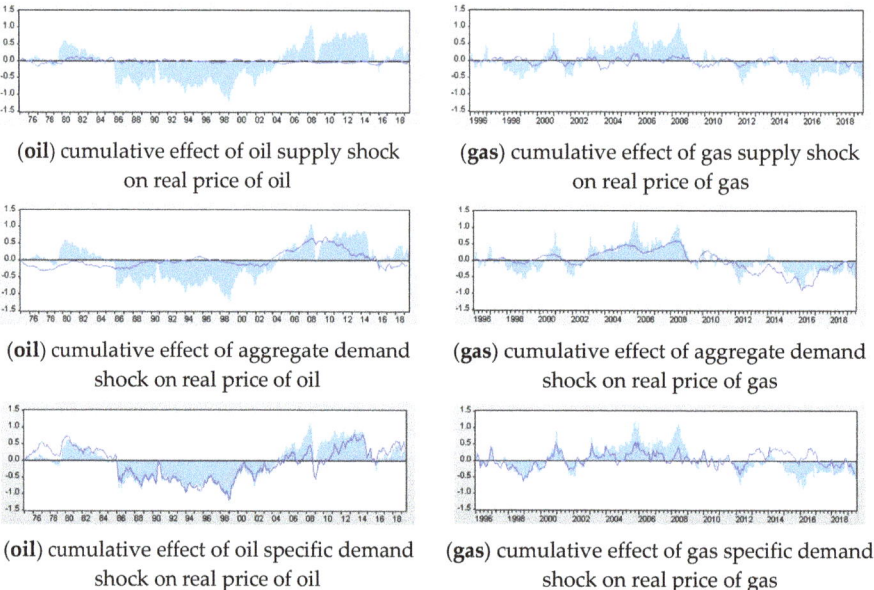

Figure 3. Historical decomposition of the real prices of crude oil and natural gas. From top to bottom, (**oil**) illustrates the cumulative effect of oil supply, aggregate demand, and oil market-specific demand shocks on the real price of crude oil. From top to bottom, (**gas**) illustrates the cumulative effect of gas supply, aggregate demand, and gas market-specific demand shocks on the real price of natural gas. Note: Estimates derived from Equations (1)–(4).

Regarding the effect of an aggregate demand expansion on real global economic activity, we can interpret from the graph that a positive aggregate demand shock caused a transitory but substantial increase that reached its maximum in period 2, followed by a gradual decline until the 8th month, and an increase again after 9 months. At the same time, a one-standard-deviation aggregate demand shock caused a very stable and statistically significant increase in the real price of oil.

The series of graphs showing the "(oil) response to an oil-specific demand shock" reveal that oil-specific demand increases had an immediate, large, persistent, and positive effect on the real price of crude oil in the first 2 months, which thereafter gradually declined but remained positive and highly statistically significant. At the same time, these shocks also triggered a statistically significant increase in real economic activity in the first 3 months, followed by a decrease that remained statistically significant until the 9th month. This compares to the results from Kilian [6], in which the statistical significance lasted until the 12th month. Regarding this difference, we estimate that the influence of specific demand shocks on global economic activity is shorter than it was before 2009, which provides evidence for the phenomenon that, as the new economic environment changes rapidly, the economic cycle shortens.

To summarize the crude oil results, aggregate demand shocks had a positive effect on both real global economic activity and the real price of crude oil. Precautionary demand shocks caused a relatively longer and positive effect on the real price of oil, and led to small, transitory increases in real economic activity. These results are generally the same as those reported by Kilian [6], except that Kilian's work showed that oil supply shocks caused a partially statistically significant and small decline in the real price of oil, and caused an extremely small increase in real economic activity, while our results showed that the shocks on the real price and the real economic activity are statistically insignificant.

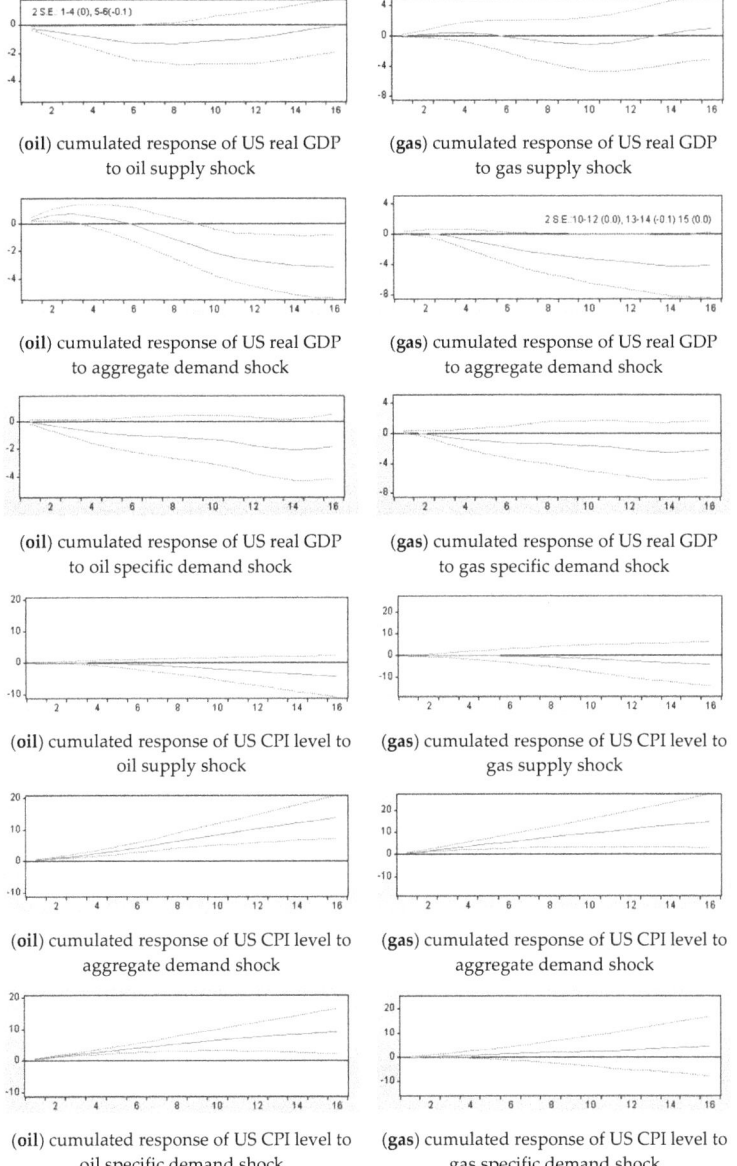

Figure 4. The responses of US real GDP growth and CPI inflation to each structural shock. **Note:** The plots show the cumulative responses, as estimated from the regression models.

Second, the empirical results of natural gas are as follows: Concerning natural gas supply shocks, positive supply shocks where production rose caused small but sharp increases, followed by a return to its initial level within the first quarter. At the same time, a one-standard-deviation natural gas supply shock caused a partially statistically significant decline in the real price of natural gas in the first 3 months, and caused a negligible increase in real economic activity.

Regarding the effect of an aggregate demand expansion on real global economic activity, we can interpret from the graph that a positive aggregate demand shock caused a transitory and highly significant increase that reached its maximum level in the 2nd month, followed by a gradual decline until the 8th month, and an increase again after 9 months. This result is similar to the result of the crude oil model. Aggregate demand shocks also caused a very stable and statistically significant increase in the real price of natural gas.

The last natural gas graph reveals that gas-specific demand increases have an immediate, noticeable, stable, and statistically significant positive effect on the real price of natural gas that gradually declines after the 3rd month but remains positive. At the same time, these shocks also triggered extremely small increases in real economic activity, followed by declines and then small increases that were statistically insignificant by the 10th month. This result suggests that precautionary demand shocks to natural gas hardly affected real economic activity.

To summarize the natural gas results, aggregate demand shocks had a positive effect on real global economic activity and the real price of natural gas. Precautionary demand shocks caused a relatively shorter, positive effect on the real price of natural gas.

Comparing the results between crude oil and natural gas reveals that they are similar, with minor differences. The major distinction relates to the different effects on real economic activity that result from precautionary demand shocks to crude oil versus natural gas.

3.1.3. The Cumulative Effect of Oil and Gas Demand and Supply Shocks on the Real Prices of Oil and Gas

According to the results of a historical decomposition of the data sets, the respective cumulative contributions of demand and supply shocks to the real prices of crude oil and natural gas are plotted in Figure 3.

Figure 3 reveals the respective cumulative contributions of oil and gas demand and supply shocks to the real prices of oil and gas.

First, the empirical results for crude oil are as follows: The top graph implies that real oil supply shocks have historically made relatively small contributions to the real price of crude oil. The biggest contributions have been oil market-specific demand shocks. While aggregate demand shocks led to long swings in real global economic activity, oil market-specific demand shocks were the primary reasons for relatively sharp and defined increases or decreases in the real price of oil. This result is consistent with the results investigated above, and provides further evidence for the proposition that precautionary demand shocks cause relatively immediate increases in real prices. Furthermore, the shocks cause relatively small but stable increases in real economic activity.

To summarize these results for crude oil, precautionary demand shocks made the greatest contributions to the real price of oil. Our results for crude oil are approximately similar to Kilian [6], but one concern is that, during the period 1985–2000, Kilian's study showed a negative cumulative effect of an aggregate demand shock on the real price of crude oil, while the effect in our study was extremely close to zero.

Second, the empirical results for natural gas are as follows: The first graph shows that real gas supply shocks have historically made relatively small contributions to the real price of natural gas. The largest contributions were due to global aggregate demand shocks and specific demand shocks in the oil market. However, aggregate demand shocks displayed a long-term and relatively smooth movement in the real price of natural gas. In contrast, gas market-specific demand shocks accounted for the sharply defined increases and decreases. This result implies the same suggestion that we already stated above regarding crude oil.

To summarize the natural gas results, the cumulative contribution of precautionary and aggregate demand shocks accounted for a large proportion of the real price of natural gas. From these empirical results for natural gas, attention should be paid to the fact that, during the years of financial crises in 1998 and 2008, the degree of the cumulative effect of the precautionary demand shocks was different;

specifically, the cumulative contribution of the oil-specific demand shock to the real price of oil is larger than the cumulative contribution of the gas-specific demand shock to the real price of gas. This issue remains open for investigation in the future.

Comparing these two results, the major difference between oil and gas was the cumulative effect of precautionary demand shocks on their real prices.

3.2. Effects on the US Economy

Based on estimations from Equations (5)–(8), Figure 4 plots the responses of US real GDP and CPI to each of the three shocks defined previously.

The results in Figure 4 reveal important differences in how oil and gas demand and supply shocks that underlie the real prices of oil and gas affect US macroeconomic aggregates such as US real GDP growth and CPI inflation.

First, the empirical results for crude oil are as follows: The graphs of the (oil)-cumulative response of US real GDP/CPI levels to an oil supply shock imply that oil supply shocks created a partially statistically significant decrease in US real GDP from its initial level and a rise in the 3rd year. The effect on US real GDP was negative for all 16 quarters; however, the two standard error bands were statistically significant for only half a year, compared to the two years reported in Kilian [6]. In contrast, the corresponding effects on the CPI level were mostly statistically insignificant, and displayed a declining tendency.

The graphs of the (oil)-cumulated response of the US real GDP/CPI levels to an aggregate demand shock indicate that unanticipated aggregate demand shocks increased US real GDP statistically significantly in the first year, and then caused the GDP level to decline below its initial level in the 2nd year, with the decrease becoming statistically significant again after the 3rd year. In contrast, the corresponding effects on the level of CPI indicated a persistent, gradual increase that was statistically significant at all time horizons observed, whereas in Kilian's results it became statistically significant only after the 3rd quarter [6].

Finally, unanticipated oil-specific demand shocks lowered US real GDP gradually, and the reduction was statistically insignificant at all observed time horizons. In contrast with this result, Kilian [6] showed that the decline reached its maximum in the 3rd year, and implied statistical significance in the 3rd year of under one standard error band. It should be noted that Kilian's results were also statistically insignificant under two standard error bands. Meanwhile, these shocks led to a sustained, gradual, and statistically significant increase in the US CPI level, which is consistent with Kilian's results [6].

To summarize the results for crude oil, our first result was that oil supply shocks led to transitory reductions in the US real GDP level, and barely affected the CPI level. Second, aggregate demand shocks had positive effects on US GDP growth during the first year and stimulated immediate increases in the CPI level, which is a different result from Kilian [6]. Kilian found that aggregate demand shocks caused delays in the rise of the CPI level for 3 quarters. Third, oil-specific demand shocks led to increases in the CPI inflation level and barely affected the US real GDP level.

Second, the empirical results for natural gas are as follows: The graphs of the (gas)-cumulative response of US real GDP/CPI level to a gas supply shock imply that the responses of the US real GDP and CPI levels to gas supply shocks are relatively small and statistically insignificant.

The graphs of the (gas)-cumulative response of US real GDP/CPI levels to an aggregate demand shock reveal that unanticipated aggregate demand shocks caused extremely small, statistically insignificant increases in the US real GDP level within 2 quarters, and then caused the real GDP level to decline below its initial level after the 3rd quarter, with the decrease becoming partially statistically significant after the 3rd year. In contrast, the corresponding effect on the level of CPI showed a sustained, gradual increase, and was statistically significant for all 16 quarters.

The graphs of the (gas)-cumulative response of US real GDP/CPI levels to a gas-specific demand shock indicate that unanticipated gas-specific demand shocks reduced the US real GDP level gradually, slightly, and statistically insignificantly. In contrast, these shocks led to sustained, gradual, but statistically insignificant increases in the US CPI level.

To summarize the results for natural gas, our first result was that gas supply shocks barely affected the level of US real GDP or CPI. Second, aggregate demand shocks caused a gradual decline in US real GDP growth that was partially statistically significant, but stimulated an immediate and statistically significant increase in the CPI level. Third, gas-specific demand shocks hardly affected the US real GDP and CPI inflation levels.

To compare these two results, the major difference was that precautionary demand shocks in crude oil and natural gas had different effects on CPI inflation. Oil market-specific demand shocks tended to statistically significantly increase the CPI inflation level greatly; in contrast, gas market-specific demand shocks led to a small and statistically insignificant increase in the CPI inflation level. Another point is that aggregate demand shocks in the oil market increase and then lower the US real GDP level significantly for the most part, whereas aggregate demand shocks in the gas market led to insignificant decreases in the US real GDP level for the most part.

4. Conclusions

This paper investigated the response of US macroeconomic aggregates to price shocks in crude oil and natural gas markets using a two-step approach: firstly decomposing the price shocks to identify the structural shocks using a structural VAR model, and secondly, showing the response patterns of macroeconomic aggregates corresponding to each structural shock using a regression model.

Our main findings in this study are threefold. First, we found that there was a difference in the effects on real economic activity due to precautionary demand shocks from crude oil versus natural gas. Second, the cumulative contribution of oil-specific demand shocks to real oil prices is larger than the cumulative contribution of gas-specific demand shocks to real gas prices. Third, precautionary demand shocks had different effects on CPI inflation, depending on whether they were due to crude oil or natural gas. Therefore, economists, policymakers, and investors must decompose real prices of these important energy sources that have such significance in the economy and our daily lives. Finally, the most noteworthy of our findings is that oil market-specific demand shocks lead to statistically significant increases in the US CPI level, however, gas market-specific demand shocks barely affect the US CPI level. This result implies policymakers should react differently to gas-specific demand shocks and oil-specific demand shocks.

The similarities and differences between our empirical results and those of Kilian [6] can be summarized as follows. Firstly, comparing the results which showed the responses of global oil/gas production, real economic activity, and the real price of oil/gas to structural innovation reveal that they are alike with minor differences. The major distinction is the different effects on real economic activity that result from precautionary demand shocks to crude oil versus natural gas. Another point is that the statistically significant lasting period of influence from oil-specific demand shocks on global economic activity were shorter than they were prior to 2009, as the results from Kilian [6] showed. Secondly, comparing the results of the historical decomposition of real prices, our results for crude oil are approximately similar to Kilian, but one concern is that, during the period 1985–2000, Kilian's study showed a negative cumulative effect of an aggregate demand shock on the real price of crude oil, while the effect in our study was extremely close to zero [6]. Thirdly, comparing the result of effects of respective shocks on the US macroeconomy, the empirical results are very similar with minor differences. The results of oil supply shocks on GDP showed that the two standard error bands were statistically significant for only half a year, compared to the two years reported in Kilian [6], which was explained by the one standard error band. Because of the different adoption of the standard error band, unanticipated oil-specific demand shocks led to a statistically insignificant reduction in US real GDP for all observed time horizons, whereas Kilian's results implied statistical significance in the 3rd year [6].

Compared to previous literature, namely Zhang et al. (2017) [13], the similarities and discrepancies should be noted. Because not all natural gas price shocks cause GDP to decrease and the CPI to increase, when natural gas price increases originate from increasing uncertainty of the energy market, GDP would decline, and the CPI would rise. However, when natural gas price shocks are due to exogenous supply shocks, GDP would not always decrease, while the CPI would decline, which is a different outcome from that relating to aggregate shocks and specific demand shocks.

In this paper, the empirical evidence may have some implications for economists, policymakers, and investors albeit their self-evidence. Because our empirical results showed that the response patterns vary depending on the causes of the price fluctuations, not only for the crude oil market, but also the natural gas market, economists should take this into account when they construct their models for researching about oil/gas price shocks. Regarding the policy-making implications, especially for government monetary authorities such as central banks, there is a need to identify the underlying determinants that drive the price increases to make appropriate policy. For example, when the oil or gas price increase stems from an exogenous supply shock, there is almost no need to undertake any countermeasures because these kinds of shocks have a minimal effect on the macroeconomic aggregates. In contrast, with supply shocks, in which the price increases because of surging uncertainty in the oil market, thereby stimulating an increase in precautionary demand, monetary authorities should consider whether countermeasures should be undertaken to retard ongoing inflation; there is almost no need to take such countermeasures into account in the case of the gas market. Another implication concerns aggregate demand shocks: policymakers should distinguish where the aggregate demand shocks originated i.e., the crude oil market or the natural gas market. If the source of the shock lies in the oil market rather than the gas market, the corresponding responses should take into consideration prevention of a reduction of the GDP level over the long-run. As for the long-term investment implications, firstly, the investors must distinguish the origins of the oil/gas price fluctuation because the response pattern of each shock is different and requires distinct investment adjustments. Secondly, investors should pay attention to global aggregate demand shocks, and the specific demand shock rather than supply shocks, when they reconsider their investment structure, because precautionary demands shocks make a greater contribution to real price fluctuations than exogenous supply shocks. Because of developing environmental awareness, the demand for natural gas is gradually increasing and growing in importance in energy markets, thus implying that investors should take natural gas as an alternative investment choice for their portfolio adjustment when turbulence occurs in crude oil markets.

Despite our findings, this investigation has several limitations. First, the range of the dataset of natural gas prices is from 1994, which makes it impossible to compare the cumulative effect of demand and supply shocks between crude oil and natural gas during the period 1973–1993. Second, we adopted the Kilian Index (Kilian [6]); however, whether the Kilian Index is an adequate reflection of overall economic activity after 2009 is beyond the scope of this study. Because the driving forces and factors which influence global economic activity may have changed after 2009, there may now be a more appropriate index. Third, as mentioned above, during the period 1985–2000, Kilian's study showed a negative cumulative effect of an aggregate demand shock on the real price of crude oil, while the effect in our study was extremely close to zero. The proper explanation for this outcome remains unknown and requires further research.

Therefore, in our future research, first, we plan to identify a more appropriate index to reflect overall global economic activity, and compare the results with those based on the Kilian Index. Second, as noted in Section 3.1.3, the cumulative contribution of an oil-specific demand shock to real prices is larger than the cumulative contribution of a gas-specific demand shock. Thus, we intend to identify the underlying determinants or causes of this issue. Third, in order to effectively apply our results to the policy-making process and inform appropriate monetary policy, we propose to extend the results by a machine learning approach to forecast price fluctuations corresponding to various kinds of uncertainty conditions, namely financial crises, wars, emergencies, disasters, etc. Fourth, the approach of Kilian [6], followed herein, did not include analysis of cointegration; thus, we plan to expand the analysis within the framework of cointegration. Finally, assessing the effects of oil/gas price shocks on other major countries' macroeconomic aggregates is also worthy of extending the current investigation in future research.

Author Contributions: Investigation, J.S.; writing—original draft preparation, J.S.; writing—review and editing, S.H.; project administration, S.H.; funding acquisition, S.H. All authors have read and agreed to the published version of the manuscript

Funding: This work was supported by JSPS KAKENHI Grant Number (A) 17H00983.

Acknowledgments: We are grateful to four anonymous referees for their helpful comments and suggestions.

Conflicts of Interest: The authors declare no conflict of interest.

Appendix A

We plotted the Kilian index according to the data from the Lutz Kilian website, which showed the monthly index of the global real economic activity. This figure shows that fluctuations in global real economic activity are consistent with the anecdotal evidence. It is commonly accepted that when war breaks out and crises or threats occur, trade uncertainty rises and these events damage global economic activity.

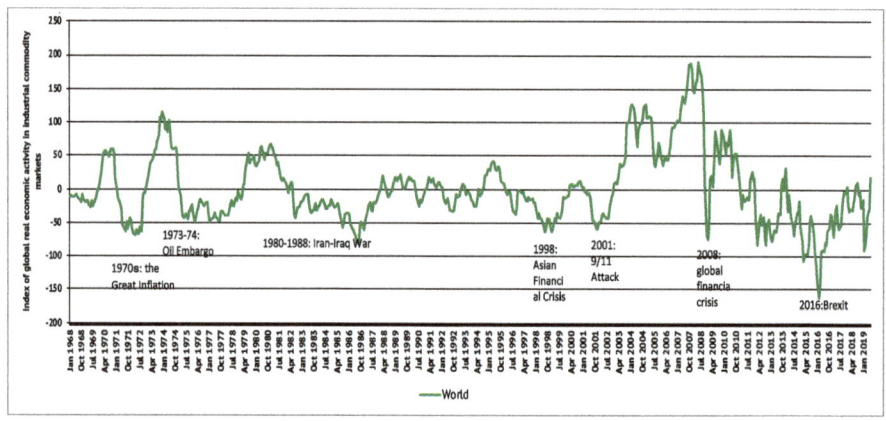

Figure A1. The Kilian Index. **Note** This figure illustrates time-changing trend of the Kilian Index [15] from Jan. 1968 to Jun. 2019.

Appendix B

The results of the SVAR stability condition check are shown below.

Table A1. The results of SVAR stability Condition Check.

Endogenous Variables: $\Delta oprod_t$ l_rea_t l_rpo_t		Endogenous Variables: $\Delta gprod_t$ l_rea_t l_rpg_t	
Exogenous variables: C		Exogenous variables: C	
Lag specification: 24 periods		Lag specification: 24 periods	
Root	Modulus	Root	Modulus
0.9893	0.9893	0.9733 − 0.0081 i	0.9734
0.9757	0.9757	0.9733 + 0.0081 i	0.9734
...	...	...	...
0.2045 − 0.5077 i	0.5473	−0.1121 + 0.2403 i	0.2652
0.2046	0.2046	−0.1121 − 0.2403 i	0.2652
No root lies outside the unit circle. VAR satisfies the stability condition.		No root lies outside the unit circle. VAR satisfies the stability condition.	
Note: $i = \sqrt{-1}$.			

References

1. International Energy Agency. World Energy Outlook 2019. 2019. Available online: https://www.iea.org/weo/ (accessed on 3 May 2020).
2. International Energy Agency. Gas-Fuels &Technologies. Available online: https://www.iea.org/fuels-and-technologies/gas (accessed on 3 May 2020).
3. Hamilton, J.D. Oil and the Macroeconomy since World War II. *J. Political Econ.* **1983**, *91*, 228–248. [CrossRef]
4. Hamilton, J.D. This is what happened to the oil price-macroeconomy relationship. *J. Monet. Econ.* **1996**, *38*, 215–220. [CrossRef]
5. Hooker, M.A. What happened to the oil price-macroeconomy relationship? *J. Monet. Econ.* **1996**, *38*, 195–213. [CrossRef]
6. Kilian, L. Not all oil price shocks are alike: Disentangling demand and supply shocks in the crude oil market. *Am. Econ. Rev.* **2009**, *99*, 1053–1069. [CrossRef]
7. Kling, J.L. Oil price shocks and stock-market behavior. *J. Portf. Manag.* **1985**, *12*, 34–39. [CrossRef]
8. Kilian, L.; Park, C. The impact of oil price shocks on the U.S. stock market. *Int. Econ. Rev.* **2009**, *50*, 1267–1287. [CrossRef]
9. Oladosu, G.A.; Leiby, P.N.; Bowman, D.C.; Uría-Martínez, R.; Johnson, M.M. Impacts of oil price shocks on the United States economy: A meta-analysis of the oil price elasticity of GDP for net oil-importing economies. *Energy Policy* **2018**, *115*, 523–544. [CrossRef]
10. Jan van de Ven, D.; Fouquet, R. Historical energy price shocks and their changing effects on the economy. *Energy Econ.* **2017**, *62*, 204–216. [CrossRef]
11. Ju, K.; Su, B.; Zhou, D.; Wu, J.; Liu, L. Macroeconomic performance of oil price shocks: Outlier evidence from nineteen major oil-related countries/regions. *Energy Econ.* **2016**, *60*, 325–332. [CrossRef]
12. Nick, S.; Thoenes, S. What drives natural gas prices?—A structural VAR approach. *Energy Econ.* **2014**, *45*, 517–527. [CrossRef]
13. Zhang, W.; Yang, J.; Zhang, Z.; Shackman, J. Natural gas price effects in China based on the CGE model. *J. Clean. Prod.* **2017**, *147*, 497–505. [CrossRef]
14. Jadidzadeh, A.; Serletis, A. How does the U.S. natural gas market react to demand and supply shocks in the crude oil market? *Energy Econ.* **2017**, *63*, 66–74. [CrossRef]
15. Kilian Index. Lutz Kilian Homepage. Available online: https://sites.google.com/site/lkilian2019/research/data-sets (accessed on 3 May 2020).
16. Sim, N.; Zhou, H. Oil prices, US stock return, and the dependence between their quantiles. *J. Bank. Financ.* **2015**, *55*, 1–8. [CrossRef]
17. Yoshizaki, Y.; Hamori, S. On the influence of oil price shocks on economic activity, inflation, and exchange rates. *Int. J. Financ. Res.* **2013**, *4*, 33–41. [CrossRef]
18. Ahmed, H.J.A.; Wadud, I.K.M.M. Role of oil price shocks on macroeconomic activities: An SVAR approach to the Malaysian economy and monetary responses. *Energy Policy* **2011**, *39*, 8062–8069. [CrossRef]

© 2020 by the authors. Licensee MDPI, Basel, Switzerland. This article is an open access article distributed under the terms and conditions of the Creative Commons Attribution (CC BY) license (http://creativecommons.org/licenses/by/4.0/).

Article

Forecasting Crude Oil Market Crashes Using Machine Learning Technologies

Yulian Zhang and Shigeyuki Hamori *

Graduate School of Economics, Kobe University, 2-1, Rokkodai, Nada-Ku, Kobe 657-8501, Japan;
zhangyulian.kobe@gmail.com
* Correspondence: hamori@econ.kobe-u.ac.jp

Received: 24 March 2020; Accepted: 8 May 2020; Published: 13 May 2020

Abstract: To the best of our knowledge, this study provides new insight into the forecasting of crude oil futures price crashes in America, employing a moving window. One is the fixed-length window and the other is the expanding-length window, which has never been reported in the past. We aimed to investigate if there is any difference when historical data are discarded. As the explanatory variables, we adapted 13 variables to obtain two datasets, 16 explanatory variables for Dataset1 and 121 explanatory variables for Dataset2. We try to observe results from the different-sized sets of explanatory variables. Specifically, we leverage the merits of a series of machine learning techniques, which include random forests, logistic regression, support vector machines, and extreme gradient boosting (XGBoost). Finally, we employ the evaluation metrics that are broadly used to assess the discriminatory power of imbalanced datasets. Our results indicate that we should occasionally discard distant historical data, and that XGBoost outperforms the other employed approaches, achieving a detection rate as high as 86% using the fixed-length moving window for Dataset2.

Keywords: oil futures prices crashes; foresting; random forests; logistical regression; support vector machines; extreme gradient boosting; moving window

---

1. Introduction

The consumption of energy commodities is becoming a crucial issue, along with modernization and technological development. Crude oil plays a pivotal role in economic growth as a major energy source. The share of America's crude oil consumption is as high as 37%, which shows the crucial position held by crude oil as a component of the energy sources [1]. For energy investors, forecasting oil price crashes can help them mitigate risk and ensure proper resource investment and allocation. Saggu and Anukoonwattaka indicate that economic growth is at significant risk from commodity price crashes across Asia-Pacific's least developed countries and landlocked developing countries [2]. Research has indicated that a crisis starts from a single economy of a large enough size and generates turbulence in other countries [3].

The beginning of a recession manifests itself through asset price drops in G-7 countries [4]. The EU allowances price drop can be justified by an economic recession [5]. By forecasting oil price crashes, we can infer recessions and develop an early warning system (EWS) for policymakers, and they can perform relevant actions to curtail the contagion crisis, or preempt an economic crisis or recession. The financial crisis of 2008 rekindled interest in EWSs. An EWS can help policymakers manage economic complexity and take precautionary actions to lower risks that can cause a crisis. An EWS can help reduce economic losses by providing information that allows individuals and communities to protect their property.

Many studies have been conducted on crude oil price forecasting [6–8]. Investigations have been conducted on oil price prediction using machine learning, along with the development and application of machine learning and deep learning [9–14]. However, what we want to know is whether there will be a huge drop in crude oil futures prices, rather than the specific numbers. Crash and crisis forecasting are significant and may be flexibly applied to finance, banking, business, and other fields. This is why crisis prediction and financial contagion have become popular topics in academic investigation in recent years, especially after the global crisis of 2008, which had a significant impact on various industries and countries. Studies in the past have provided important information on predicting financial crises [3,15] by applying machine learning. Nevertheless, there are still few studies predicting oil futures price crashes. Additionally, we want to fill the gap of crude oil futures price crash forecasting in the crude oil market. Oil price crash forecasting can give the early warning information to investors and policymakers, so they can do some precautionary action to reduce the loss.

We define the crisis based on [3,15]. Our investigation contributes to the current literature by analyzing the predictive performance of a series of state-of-the-art statistical machine learning methods, including random forest, logistic regression, support vector machine, and extreme gradient boosting (XGBoost) algorithms, in the classification problem of crude oil futures price market crash detection in America, covering the period from 1990 to 2019. To the best of our knowledge, this study provides new insights into forecasting oil price crashes, employing a moving window that has never been reported in the past. Furthermore, we also consider the previous data in the moving window algorithm, and the experimental work presented here provides a novel result that discarding previous data can achieve better performance. We develop an exhaustive experimental performance evaluation of our algorithms using a test dataset of the daily returns performance of the 25-year oil price market.

Finally, we find that the fixed-length window provides a better result than the expanding window, which indicates that we should occasionally abandon distant historical data, and XGBoost outperforms the other techniques in this study, achieving a detection rate as high as 86% using the fixed-length moving window for Dataset2.

The remainder of this paper is organized as follows. Section 2 reviews the previous literature related to this study. We provide a brief explanation of the machine learning algorithms and the moving window utilized in Section 3. In Section 4, we elaborate on the dataset implemented to perform the system. In Section 5, we present an experimental evaluation and provide empirical results. We evaluate the empirical findings based on a series of assessments, computed on a test dataset sample spanning a long period. Finally, we present some concluding remarks in Section 6.

2. Literature Review

The topic of oil price forecasting has been extensively studied [6–8]. High- and low-inventory variables have been used to forecast short-run crude oil prices [16]. The results show that global non-oil industrial commodity prices are the most successful predictors of oil prices [6]. Moshiri and Foroutan [17] developed a nonlinear model to forecast crude oil futures prices.

With the development of machine learning and deep learning, an increasing number of academic studies have attempted to research this problem using a machine learning algorithm [9–13]. Wen et al. [9] make crude oil price forecasting based on support vector machines. An empirical model decomposition based neural network ensemble learning paradigm is proposed for oil spot price forecasting [10]. An improved oil price forecast model that uses a support vector machine was developed [18]. Some researchers forecast the crude oil price using machine learning [12] and the deep learning ensemble approach [14]. Other researchers use XGBoost [13] and wavelet decomposition and neural network model [11] to forecast crude oil prices. However, compared with the specific figures of crude oil prices, investors and practitioners are more concerned about the drops or crashes of crude oil prices.

An analytic network process model is applied in forecasting the crisis [19]. There are many papers on predicting financial crises [3,15,20] that apply machine learning. Lean et al. [20] employ the general regression neural networks to predict the currency crisis upon the disastrous 1997–1998 currency crisis experience. Based on a multinomial logit model, Bussiere and Fratzscher [15] develop a new EWS model to forecast financial crises. Chatzis et al. [3] use deep and statistical machine learning techniques (classification trees, support vector machines, random forests, neural networks, extreme gradient boosting, and deep neural networks) to predict stock market crisis events. They find that forecasting accuracy can be improved by adding oil variables to the traditional predictors of excess stock returns of the S&P 500 index. However, investigations into oil price crashes are still very inadequate. The experimental work presented here provides one of the first investigations into how to forecast crashes of crude oil future prices using machine learning.

Chatzis et al. [3] detect a crisis event based on the number of stock market negative coexceedances (less than 1% percentile of the empirical distribution). Due to the imbalanced dataset, and because we want to obtain more crash events, we consider crash events in less than 2.5% percentile of the empirical distribution [15].

A new EWS model for predicting financial crises was developed based on a multinomial logit model in [15], and the model correctly predicts the majority of crises in emerging markets. The 2008 financial crisis brought huge losses to the financial industry, as well as society as a whole, and people gained renewed awareness of the importance of an EWS. Davis and Karim [21] assess and suggest that the logit is the most beneficial approach to a global EWS, along with signal extraction for country-specific EWS for banking crises on a comprehensive common dataset. Babecky et al. [22] account for model uncertainty using Bayesian model averaging, identifying early warning indicators of crises specific to developed economies. The study indicates that the dynamic Bayesian network models can offer precise early-warnings compared with the logit and signal-extraction methods [23].

Studies on EWSs for financial crises employing machine learning techniques are still limited. This paper develops an EWS that includes artificial neural networks, decision trees, and a logistic regression model [24]. Geng et al. [25] use data mining techniques to build financial distress warning models and find that the performance of neural networks is better than decision trees, support vector machines, and multiple classifiers. Finally, a total of six deep and statistical machine learning techniques are used to develop the EWS and forecast a stock market crisis [3].

In this paper, we performed the forecasting model using random forests (hereafter RF) [26], logistical regression (hereafter LogR) [27], support vector machines (hereafter SVMs) [28], and extreme gradient boosting (hereafter XGBoost). RF is applied in gene expression analysis [29], protein–protein interactions [30], risk assessment [31], and image classification [32], among others. Ohlson [27] uses LogR for the first time to predict corporate bankruptcy based on financial statement data. LogR and NN (neural networks) are employed to predict the bond rating [33]. SVMs are widely used in credit rating systems in many investigations [34,35]. XGBoost is an important and efficient implementation of the gradient boosting framework in the classification methodology of [36,37].

In terms of oil price crashes forecasting, there are many methods that can be employed. It includes nearest neighbors [38], NN (neural network), which is broadly applied in credit rating classification [39,40], and LSTM (long short-term memory), it is employed in forecasting the volatility of stock price index [41]. In the future, in addition to adding other methods, we can also consider improve the existing approach. Zhong and Enke [42] use the nature dimensionality reduction technique, means principal component analysis (PCA), and the artificial neural networks (ANNs), yielding improved outcomes.

3. Model Development

We utilized the following models: RF [26], LogR [27], SVMs [28], and XGBoost. Then, we presented a detailed interpretation of the development and parameter tuning of each machine learning methodological framework.

We used the machine learning models above, and we also considered two patterns of moving window corresponding to Section 4, aimed to investigate if there is any difference when historical data are discarded. One is the fixed-length window and the other is the expanding window.

In terms of the fixed-length window, for daily data, we considered 1000 days (almost five years) as the window length, and the first training data was the dataset of the first 1000 records starting from the first day (covering the period 18 June 1990 to 28 September 1994). We considered the 1001st piece of data as the first piece of daily test data. For every subsequent piece of test data (one day in the examined period), we incorporated a new record as the training data, and to keep the window length fixed, we rejected the first observation at the same time. Additionally, for the last observation of daily data in the sample (i.e., 18 December 2019), the empirical distribution of returns ended with the period 23 October 2015 to 17 December 2019. As the daily dataset had 7146 data points, we could obtain 6146 results of test data daily.

As for the expanding window, for daily data, we considered the first 1000 records as the first training data to test the 1001st observation (the first piece of test data), such as the fixed-length window. However, for each subsequent observation (one day in the examined period), the difference is, we incorporated a new observation again as the new training data (and did not discard the first observation). As the total return records was 7146 daily pieces of data, we considered the window length to increase from 1000 to 7145. Additionally, for the last observation of daily data in the sample (i.e., 18 December 2019), the empirical distribution of returns was based on the period from 18 May 1990 to 17 December 2019. We could also obtain 6146 results of test data daily.

3.1. Random Forest (RF)

RF is a combination of tree predictors, where every tree depends on the value of a random vector sampled independently with the same distribution for all trees in the forest [26]. This is a well-known machine learning technique and is applied in gene expression analysis [29], protein–protein interactions [30], risk assessment [31], and image classification [32], financial bankruptcy forecasting [43], among others. In order to implement the RF technique, there is a common package in R named randomForest, which is provided by Liaw and Wiener [44]. We tried using this package to forecast crashes as, because of the huge size of the dataset, the speed of calculation was very slow. Liaw and Wiener [45] also show that the randomForest package is not a fast implementation for high-dimensional data. A fast implementation of random forests for high-dimensional data, called rangers, was introduced. We employed the ranger package in R here [46].

The specific process of this algorithm is as follows. Here is a dataset D that consists of many features denoted by $X_1 - X_N$, and the dependent variable Y. The dependent variable here is binary, so we could develop a classification problem. As one of the hyperparameters, num.trees (here called n), we considered n as the number of decision trees that RF is expected to generate, and the group was called the Forest. The other hyperparameter mtry (here called m), which means the number of variables to possibly split at each node, and m < N. We tuned the hyperparameter using the tuneRF function, which has the aim of searching for the optimal value of m for RF. We set a specific value for n and the optimal value for m, and by testing several n values based on the model assessment, we could obtain a relatively good result for m and n. Based on this, we also tried to perform some numerical modification. We implemented the algorithm on the basis of the modification in R.

3.2. Logistic Regression (LogR)

LogR is a famous statistical approach using logistic regression to perform classification, which is usually used in differentiating binary dependent variables. Babecky et al. [22] use LogR for the first time to predict corporate bankruptcy based on financial statement data. LogR and NN (neural networks) are employed to predict the bond rating [33]. A credit scorecard system is also usually built [47]. We develop the LogR in R via the glm function, which is used to fit generalized linear models, specified by giving a symbolic account for the linear predictor and an account for the error distribution. We adjust the thresholds due to the imbalance between the crash and no crash events.

3.3. Support Vector Machines (SVMs)

SVMs [28,48] are learning machines for two group classification problems. SVMs are widely used in credit rating systems in many investigations [34,35]. In this study, we tested the SVMs using linear, radial basis function (RBF), polynomial, and sigmoid kernels. In order to choose the proper kernel, we changed the kernel by using it into the moving window and, finally, selected the polynomial kernel. We considered soft-margin SVMs. For the cost hyperparameters, C, of the SVMs (relative to the soft-margin SVMs), and the I3 hyperparameter, we employed a grid-search algorithm. We also considered the different weights of crash event and no crash event according to the proportion of the imbalanced dataset. We implemented this SVMs algorithm in R using the e1071 package, along with the grid-search functionality, in this package.

3.4. Extreme Gradient Boosting (XGBoost)

The last main approach we used in this study to predict the crude oil market turbulence is the XGBoost (extreme gradient boosting) algorithm. It is an important and efficient implementation of the gradient boosting framework in the classification methodology of [36,37]. Two algorithms are included: one is the linear model and the other is the tree-learning algorithm. It also provides multifarious objective functions, which include classification, regression, and ranking. We developed XGBoost in this study by using the XGBoost R package. Here, we employed a tree-learning model to perform a binary classification.

We selected the statistical machine learning techniques after taking the general performance and the calculation time required. In addition to the above methods, there are many other approaches that can be applied in this field. It includes the nearest neighbors [38], NN (neural network), and LSTM (long short-term memory). Especially, NN is a famous machine learning algorithm and it is broadly employed in credit rating classification filed [39,40]. A simple NN model is composed of three types of layers, which includes the input layer, hidden layer, and output payer. In the input layer, there are the all candidate variables as a high dimensional vector, and the it is transformed into a lower-dimensional potential information in the hidden layer, and then the output layer can generate the predictions by using the non-linear functions, such as a sigmoid. Even we could also consider using SGD (stochastic gradient descent), PCA (principal component analysis), and the neural network Adam to optimal the model in the future.

Following the moving window that corresponds to Section 3.2, we tested the series of entailed hyperparameters, including the maximum depth of the generated trees, the maximum number of central processing unit threads available, and the maximum number of iterations (for classification, it is similar to the number of trees to grow). On account of the nature of the dependent variable, we employed the "binary: logistic" objective function to train the model. For a better understanding of the learning process, we used an evaluation metric of the area under the receiver operating characteristic (AUROC) to obtain a value for the area under the curve (AUC). The AUROC curve is a performance measure for classification problems at various threshold settings, and the AUC shows the capability of distinguishing between classes. In the analysis, the value of AUC usually varies between 0.5 and 1, and a higher AUC value demonstrated that the model could better distinguish the crash and no crash events. An AUC value of above 0.9 (the result of the last prediction in the moving window) indicates a very good performance of the algorithm in this study.

4. Data Collection and Processing

We considered various relative variables to forecast crashes in the crude oil markets. On the other hand, we also considered whether the variables could provide an adequate data sample. Based on the criteria, we excluded those variables with a relatively short time series. We used the futures prices of crude oil from 13 June 1986 to 18 December 2019, and the futures prices of natural gas and gold, as well as the futures prices of agricultural commodities, which included rough rice, wheat, corn, sugar, cocoa, and canola. We also considered data from the VIX index, the S&P 500 stock price index, and the yield of the USA 10-year bond from 8 June 1990 to 18 December 2019 for daily data. The data were sourced from Bloomberg and Datastream, and detailed information can be seen in Table 1.

Table 1. Exploratory variables in the model.

Variable	Data	Data Source
Crude Oil	WTI Crude Oil futures prices	
Natural Gas	Henry Hub Natural Gas futures prices	
VIX Index	Chicago Board Options Exchange Volatility Index	
Bond	USA 10-year bond yield	Bloomberg
Stock	S&P 500	
Gold	Gold futures prices	
Rough Rice	Rough Rice futures prices	
Wheat	Wheat futures prices	Datastream
Corn	Corn futures prices	
Sugar	Sugar futures prices	
Cocoa	Cocoa futures prices	Bloomberg
Canola	Canola futures prices	

Note: WTI is the West Texas intermediate. VIX is the Volatility Index.

As a result, we obtained a 30-year period for daily data, which included many crude oil futures price crashes. In this way, we filled the datasets with 7146 records for daily use. They provide the capability for modeling contagion dynamics, financial markets, and commodity market interdependencies over time.

In [3], the crisis event is detected based on an empirical distribution that is less than the 1% percentile. In order to obtain more crash event samples, we defined the crash event as less than the 2.5% percentile of the empirical distribution due to the imbalanced dataset, following [15]. That is, we defined the "crash event" ($Crash_{OIL_t}$) at each working day. If the log return (hereafter return) of the crude oil futures prices is less than the 2.5% percentile of the empirical distribution of the return, it can be detected. We could obtain a binary variable of the crude oil futures price crashes as follows:

$$Crash_{OIL_t} = \begin{cases} 1 & \text{if } R_{OIL_t} < E(R_{OIL_t}|I_{OIL_{t-1}}) - 2\sqrt{Var(R_{OIL_t}|I_{OIL_{t-1}})} \\ 0 & \text{otherwise} \end{cases}, \qquad (1)$$

where R_{OIL_t} is the log return of crude oil futures prices, $I_{OIL_{t-1}} = \{R_{OIL_{t-1}}, R_{OIL_{t-2}} \cdots\}$ is the information set, $E(R_{OIL_t}|I_{OIL_{t-1}})$ denotes the sample mean of time t under condition $I_{OIL_{t-1}}$, and $\sqrt{Var(R_{OIL_t}|I_{OIL_{t-1}})}$ is the standard deviation of the crude oil futures prices of time t under condition $I_{OIL_{t-1}}$.

Based on this, we employed two patterns of moving window to calculate the crash events. One is the fixed-length window and the other is the expanding window.

In terms of the fixed-length window, for daily data, we considered 1000 days (almost five years) as the window length and calculated the initial empirical distribution of returns (starting from the first day) based on the crude oil futures prices returns of the first 1000 observations (covering the period from 16 June 1986 to 7 June 1990). We could detect if the 1001st is a crash event according the first empirical distribution of daily returns. For each subsequent data point (day in the examined period), we recalculated the empirical distribution of returns and incorporated a new record, and in order to keep the window length fixed, we rejected the first observation at the same time. The crash event can be identified when the return was below the 2.5% percentile of the new empirical distribution. In addition, for the last observation of daily data in the sample (i.e., 18 December 2019), the empirical distribution of returns ended with the period 23 October 2015 to 17 December 2019.

For the expanding window, for daily data, we could calculate the first empirical distribution based on the first 1000 records to identify the 1001st observation, similar to the fixed-length window. However, for each subsequent observation (day in the examined period), the difference is that we calculated the empirical distribution of returns and incorporated another new observation (we did not discard the first observation). As the total daily return records total 8152, we considered the window length to increase from 1000 to 8151. In addition, for the last observation of daily data in the sample (i.e., 18 December 2019), the empirical distribution of returns was based on the period of 16 June 1986 to 17 December 2019.

Ultimately, we obtained 7152 daily crash records (binary variables). We created the extended variables to take Lag1 to Lag5 of return (we elaborated these in Table 2), so we omitted the first six crash records.

Table 2. Extended variables. Returns are the log returns.

Variable	Transformation
rash (binary variables)	Lag1–Lag5, L5D
IX Index	Lag1–Lag5
Crude Oil futures prices	
Natural Gas futures prices	
USA 10-year bond yield	
S&P 500	
Gold futures prices	Lag1–Lag5, Lag1–Lag5 of returns
Rough Rice futures prices	
Wheat futures prices	
Corn futures prices	
Sugar futures prices	
Cocoa futures prices	
Canola futures prices	

Note: L5D is the average of the crash events for the last five working days.

Figure 1 shows the number of crude oil market crashes for daily data during the selected 30-year period. We could see several severe crashes in the daily and weekly frequencies, especially in 1990, 1996, 2008, and 2015. The most severe crashes were in the global crisis of 2008, and the number of crashes was up to 35.

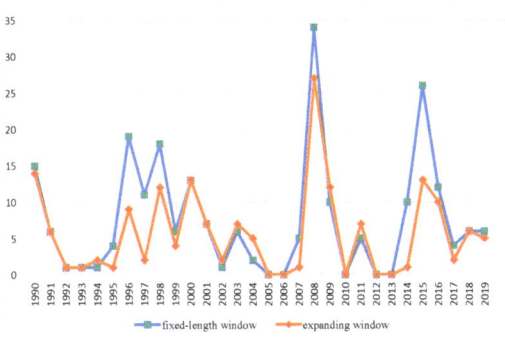

Figure 1. Number of crude oil market crashes for daily data (exceedance less than 2.5% percentile of the empirical distribution). Note: The fixed-length window means the length of the window was fixed at 1000 days, and the expanding window means the length of window increased from 1000 days to 8151 days. The total daily data was 8152.

On the other hand, the raw data used are shown in Table 1. In order to get the most out of the raw data and give full play to the machine learning models, we extended the raw variables and obtain the extended variables. To predict the crash events, we considered the previous data of crash events as a fairly predictive indicator, so we also considered the binary variables of crash events.

We made the following transformations based on the exploratory variables to capture the subtler dependencies and dynamics. Detailed information is displayed in Table 2.

1. We computed the lagged variables on a daily basis for each crash indicator, starting from one to five days (Lag1–Lag 5).
2. Except for the Crash variable (binary variables) and VIX index (because it is a volatility index, we did not return to it), we took the return of the other variables. Then, using (i), we calculated the lagged variables of the return variables again.
3. We calculated the average number of crash events for the last five working days (L5D), according to the binary variables of crash events on a daily basis.

We also developed two predictive indicator datasets to see how much the performance of the machine learning models improves. One had only the Crash (Lag1–Lag5) and the crude oil futures prices (Lag1–Lag 5, Lag1–Lag5 of returns) and the average of crash events for the last five working days (L5D; hereafter Dataset1), and the other had all predictive indicators (hereafter Dataset2). In Dataset1, there were 16 predictors, and in Dataset2, there were 121 predictors. For daily data, the dataset had 7146 data points, spanning from 18 June 1990 to 18 December 2019.

5. Experimental Evaluation

After we obtained the confusion matrix of the various models for machine learning, we performed an experimental evaluation procedure. In other words, we showed the performance results for the evaluation of the models covering a long period (1994–2019 for daily data), which included several crash events.

5.1. Model Validation Measures

Classification accuracy is the main criterion for evaluating the efficiency of each approach and for selecting the most robust method according to discriminatory power. We provided several metrics that are widely used to quantitatively assess the discriminatory power of each machine learning model. However, our dataset faced an issue of imbalanced amounts of the two classes, called imbalanced data. For example, in terms of the fixed length of the moving window for daily data, there were 6146 data points in total, in which only 205 were "CRASH." Even if we obtained high classification accuracy, we could not consider the model to perform well. As such, there is a risk that the assessment norm we used may misinterpret based on the skewed class distribution. For the imbalanced data, we wanted to know the capacity of predicting the minority of the dataset.

Bekkar et al. [49] presented a set of evaluation measures for model assessment over imbalanced datasets. First, they obtained the sensitivity and specificity based on the confusion matrix and considered combined measures (G-means, likelihood ratios, F-measure balanced accuracy, discriminant power, the Youden index, and the Matthews correlation coefficient). They also considered the graphical performance assessment (receiver operating characteristic (ROC) curve, area under curve, etc.). Due to the application of the moving window, we could not obtain the AUC value based on the area under the curve metrics, so we do not consider the AUC as the general metric. In a similar way, we do not use the graphical performance assessment. In machine learning, the two classes of forecasting method are assessed by the confusion matrix, such as in Table 3. The raw data represents the real value of the class label, while the column represents the classifier prediction. In the imbalanced dataset, the observations of the minority class were defined as positive, and the observations of the majority were labeled negative. Here, positive means "CRASH," negative means "NO CRASH." Following [38], we adopted the sensitivity and specificity metrics, which are defined as follows.

Using the confusion matrix given by Table 3, we could calculate the sensitivity and specificity as follows:

$$Sensitivity = \frac{TP}{TP + FN} \quad (2)$$

$$Specificity = \frac{TN}{TN + FP} \quad (3)$$

Table 3. Confusion matrix for two-class classification.

	Predicted No Crash	Predicted Crash
Actual no crash	TN	FP
Actual crash	FN	TP

Note: *TP* = True Positive, means the number of positive cases (i.e., crash) that are identified as positive correctly; *FP* = False Positive, shows the number of negative vases (i.e., no crash) that are misclassified as positive cases; *FN* = False Negative, indicates the number of positive cases that are incorrectly identified as negative cases; *TN* = True Negative, is the number of negative cases that are correctly identified as negative cases.

Sensitivity is the ratio of true positive to the sum of true positive and false negative, i.e., the proportion of actual positives that are correctly identified as such. Specificity is the ratio of true negative to the sum of true negative and false positive, i.e., the proportion of actual negatives that are correctly identified as such. Based on sensitivity and specificity, we calculated several combined evaluation measures as follows:

G-mean: The geometric mean (*G-mean*) is the product of sensitivity (accuracy on the positive examples) and specificity (accuracy on the negative examples). The metric indicates the balance between classification performances in the minority and majority classes. The *G-mean* is defined as follows:

$$G = \sqrt{Sensitivity \times Specificity} \tag{4}$$

Based on this metric, even though the negative observations are correctly classified per the model, a poor performance in the prediction of the positive examples will lead to a low *G-mean* value.

LR (+): The positive likelihood ratio (*LR (+)*) represents the ratio between the probability of predicting an example as positive when it is truly positive and the probability of the predicted example being positive when, in fact, it is not positive. We have

$$LR\,(+) = \frac{Sensitivity}{1 - Specificity} \tag{5}$$

LR (−): The negative likelihood ratio (*LR (−)*) is defined as the ratio between the probability of predicting an example as negative when it is actually positive, and the probability of predicting an example as negative when it is truly negative. It is written as:

$$LR\,(-) = \frac{1 - Sensitivity}{Specificity} \tag{6}$$

As we can see, the higher *LR (+)* and lower *LR (−)* show better performance in the positive and negative classes, respectively.

DP: Discriminant power (*DP*) is a metric that summarizes sensitivity and specificity, defined as follows:

$$DP = \frac{\sqrt{3}}{\pi}\left(\log\frac{Sensitivity}{1 - Sencitivity} + \log\frac{Specificity}{1 - Specificity}\right) \tag{7}$$

A *DP* value higher than 3 indicates that the algorithm distinguishes between positive and negative examples.

BA: The balanced accuracy (*BA*) assessment is the average of sensitivity and specificity. This metric performs equally well in either class. It holds:

$$BA = \frac{1}{2}(Sensitivity + Specificity) \tag{8}$$

In contrast, if the conventional accuracy is high only because the classifier can distinguish the majority class (i.e., 'no crash' in our study), the *BA* value will drop.

WBA: The weighted balanced accuracy (*WBA*) is the weighted average of sensitivity of 75% and a specificity of 25% on the basis of *BA* [3].

Youden index: The *Youden index* γ measures the ability of the algorithm to avoid failure. It is defined as:

$$\gamma = Sensitivity + Specificity - 1 \qquad (9)$$

Generally, a higher value of γ indicates a better ability to avoid misclassification.

We used these criteria to perform a comprehensive measurement of the discriminative power of each technique. On the basis of these outcomes, we inferred an optimal cutoff threshold of the predicted crash probabilities for each fitted model, pertaining to the optimal sensitivity and specificity metrics. The results of each evaluation are displayed in Tables 4 and 5. As we could infer from Tables 4 and 5, both in Dataset1 and Dataset2, in terms of a fixed-length window, XGBoost could lead to the highest classification accuracy, and in the case of an expanding window, LogR outperformed XGBoost.

Table 4. Model validation measures of daily data for Dataset1.

Window for Fixed-Length	RF	LogR	SVMs	XGBoost
G-mean	0.509	0.806	0.501	0.846
LR (+)	7.852	5.028	3.750	8.339
LR (−)	0.758	0.276	0.786	0.232
DP	0.560	0.695	0.374	0.858
BA	0.617	0.807	0.599	0.848
WBA	0.443	0.786	0.435	0.819
Youden	0.234	0.614	0.199	0.695
Window for Expanding	**RF**	**LogR**	**SVMs**	**XGBoost**
G-mean	0.345	0.590	0.406	0.463
LR (+)	3.717	1.837	1.699	2.998
LR (−)	0.907	0.716	0.915	0.832
DP	0.338	0.226	0.148	0.307
BA	0.545	0.606	0.538	0.578
WBA	0.334	0.536	0.361	0.405
Youden	0.090	0.212	0.076	0.155

Note: In Dataset1, there are 16 predictors. *G-mean* represents the geometric mean, *LR (+)* is the positive likelihood ratio, *LR (−)* denotes the negative likelihood ratio, *DP* denotes the discriminant power, *BA* represents the balanced accuracy, and Youden denotes the *Youden index* I3.

Table 5. Model validation measures of daily data for Dataset2.

Window for Fixed-Length	RF	LogR	SVMs	XGBoost
G-mean	0.459	0.527	0.421	0.847
LR (+)	3.766	3.690	10.116	5.124
LR (−)	0.825	0.760	0.834	0.164
DP	0.364	0.378	0.597	0.824
BA	0.582	0.610	0.581	0.847
WBA	0.403	0.456	0.381	0.855
Youden	0.165	0.220	0.163	0.695
Window for Expanding	**RF**	**LogR**	**SVMs**	**XGBoost**
G-mean	0.402	0.560	0.296	0.366
LR (+)	3.058	2.102	5.807	5.956
LR (−)	0.878	0.754	0.925	0.883
DP	0.299	0.245	0.440	0.457
BA	0.558	0.601	0.537	0.557
WBA	0.364	0.492	0.313	0.347
Youden	0.115	0.201	0.074	0.114

Note: In Dataset2, there are 121 predictors. *G-mean* represents the geometric mean, *LR (+)* is the positive likelihood ratio, *LR (−)* denotes the negative likelihood ratio, *DP* denotes the discriminant power, *BA* represents the balanced accuracy, and Youden denotes the *Youden index* I3.

5.2. Accuracy of the Generated Alarms

In conclusion, we provided a new insight into developing an early warning system for increasing awareness of an impending crash event in the crude oil futures prices market, using machine learning techniques, which included random forest, logistic regression, support vector machine, and extreme gradient boosting algorithms. We calculated an optimal oil futures price crash probability cut-off threshold for each model, and when the predictive probabilities exceed the settled threshold, an alarm will be generated.

As we can see from Tables 6 and 7, we obtained the final confusion matrix for all the evaluated models. The results of the false alarm rate and detection rate are also shown. Concentrating on the best-performing model, namely the XGBoost of the fixed-length window, by accepting a 9% false alarm rate, we succeeded in forecasting 79% of futures price crashes in the crude oil market for Dataset1. Under a false alarm rate of 17%, we could even predict 86% of crashes for Dataset2.

Table 6. Classification accuracy table for Dataset1.

RF	pred		LogR	pred		SVMs	pred		XGBoost	pred	
	Fixed-Length Window										
ture	0	1	Ture	0	1	ture	0	1	ture	0	1
0	5738	203	0	5036	905	0	5832	454	0	5378	563
1	150	55	1	48	157	1	175	65	1	43	162
Signal	FAR	DR	Signal	FAR	DR	Signal	FAR	DR	Signal	FAR	DR
Rate	3%	27%	Rate	15%	77%	Rate	7%	27%	Rate	9%	79%
	Expanding Window										
RF	pred		LogR	pred		SVMs	pred		XGBoost	pred	
ture	0	1	Ture	0	1	ture	0	1	ture	0	1
0	5801	199	0	4479	1521	0	5437	653	0	5534	466
1	128	18	1	78	68	1	119	27	1	112	34
Signal	FAR	DR	Signal	FAR	DR	Signal	FAR	DR	Signal	FAR	DR
Rate	3%	12%	Rate	25%	47%	Rate	11%	18%	Rate	8%	23%

Note: In Dataset1, there are 16 predictors. FAR represents false alarm rate, $FAR = \frac{FP}{FP+TN}$; DR represents detection rate, and $DR = \frac{TP}{TP+FN}$.

Table 7. Classification accuracy table for Dataset2.

RF	pred		LogR	pred		SVMs	pred		XGBoost	Pred	
	Fixed-Length Window										
ture	0	1	ture	0	1	ture	0	1	ture	0	1
0	5587	354	0	5454	487	0	5835	106	0	4940	1001
1	159	46	1	143	62	1	168	37	1	28	177
Signal	FAR	DR	Signal	FAR	DR	Signal	FAR	DR	Signal	FAR	DR
Rate	6%	22%	Rate	8%	30%	Rate	2%	18%	Rate	17%	86%
	Expanding Window										
RF	pred		LogR	pred		SVMs	pred		XGBoost	Pred	
ture	0	1	ture	0	1	ture	0	1	ture	0	1
0	5664	336	0	4905	1095	0	5908	92	0	5862	138
1	121	25	1	90	56	1	133	13	1	126	20
Signal	FAR	DR	Signal	FAR	DR	Signal	FAR	DR	Signal	FAR	DR
Rate	6%	17%	Rate	18%	38%	Rate	2%	9%	Rate	2%	14%

Note: In Dataset2, there were 121 predictors. FAR represents false alarm rate, $FAR = \frac{FP}{FP+TN}$; DR represents detection rate, and $DR = \frac{TP}{TP+FN}$.

6. Some Concluding Remarks

Financial crisis prediction plays a pivotal role for both practitioners and policymakers, as they can infer crashes and recessions through an early warning system, being able to perform relevant actions that curtail the contagion crisis or preempt an economic crisis or recession.

Firstly, we chose the most significant financial market indicators to help forecasting and perform changes that capture the subtler dependencies and dynamics, and defined the crash event as less than the 2.5% percentile of the empirical distribution. Then we selected the statistical machine learning techniques after taking the general performance and the calculation time required, and we developed the evaluated models by tuning the hyperparameter and obtained the optimal performance of each algorithm. We employed the evaluation metrics that are broadly used to assess the discriminatory power of a binary classifier on imbalanced datasets, finally.

Lean et al. [20] developed a general regression neural network (GRNN) currency forecasting model, and compared its performance with those of other forecasting methods. An early warning system was also developed using artificial neural networks (ANN), decision trees, and logistic regression models to predict whether a crisis happens within the upcoming 12-month period [24]. Additionally, the result shows ANN has given superior results. Chatzis et al. [3] use a series of techniques including classification trees, SVMs, RF, NN, XGBoost, and deep neural networks to forecast the stock market crisis events, and find that deep neural networks built using the MXNET library is the best forecasting approach. Our investigation concerns developing machine learning models that can be continuously retrained in moving window setup, which Chatzis et al. [3] do not challenge. Next, many previous investigations only use accuracy to measure the predictive ability of machine learning methods [20,24]. However, in terms of the imbalanced dataset, we were concerned about whether the crisis or crash event can be predicted precisely, rather than the no crisis or no crash event. Additionally, we regarded the detection rate as important.

The main novel contribution of this empirical investigation to the existing literature is that it fills the gap of futures prices forecasting in the crude oil market, and we pioneered the use of a moving window in this field. We also made a comparison where the previous data were discarded.

Our empirical results show the following:

1. Except for the LogR for Dataset2, in other machine learning algorithms, the fixed-length window shows a better performance than the expanding window. This indicates that discarding historical data is a better choice in forecasting the future.
2. XGBoost outperformed the rest of the employed approaches. It could even reach an 86% detection rate using the fixed-length moving window for Dataset2.
3. The performances of Dataset1 and Dataset2 did not differ significantly, meaning the indicators that are only about crude oil futures prices and crashes were very important.

These findings made our study much more attractive to researchers and practitioners working in petroleum-related agencies. The result of XGBoost also provided strong evidence that it could offer a good starting point for developing an early warning system.

For future work, first, we will consider feature engineering and perform feature selection to improve the accuracy of Dataset2. Chatzis et al. [3] show that deep neural networks significantly increase classification accuracy for forecasting stock market crisis events. Owing to the use of the moving window, we encountered some difficulties in coding when we wanted to implement the neural network technique. We will work hard to overcome this in future work.

Author Contributions: Investigation, Y.Z.; writing—original draft preparation, Y.Z.; writing—review and editing, S.H.; project administration, S.H.; funding acquisition, S.H. All authors have read and agreed to the published version of the manuscript.

Funding: This work was supported by JSPS KAKENHI Grant Number (A) 17H00983.

Acknowledgments: We are grateful to three anonymous referees for their helpful comments and suggestions.

Conflicts of Interest: The authors declare no conflict of interest.

References

1. *Monthly Energy Review*; U.S. Energy Information Administration. Available online: https://www.eia.gov/totalenergy/data/monthly/index.php (accessed on 10 February 2020).
2. Saggu, A.; Anukoonwattaka, W. Commodity Price Crash: Risks to Exports and Economic Growth in Asia-Pacific LDCs and LLDCs. *United Nations ESCAP Trade Insights* **2015**, *6*, 2617542.
3. Chatzis, P.S.; Siakoulis, V.; Petropoulos, A.; Stavroulakis, E.; Vlachogiannakis, N. Forecasting stock market crisis events using deep and statistical machine learning techniques. *Expert Syst. Appl.* **2018**, *112*, 353–371. [CrossRef]
4. Bluedorn, J.C.; Decreddin, J.; Terrones, M.E. Do asset price drops foreshadow recessions? *Int. J. Forecast.* **2016**, *32*, 518–526. [CrossRef]
5. Koch, N.; Fuss, S.; Grosjean, G.; Edenhofer, O. Causes of the EU ETS price drop: Recession, CDM, renewable policies or a bit of everything?–New evidence. *Energy Policy* **2014**, *73*, 676–685. [CrossRef]
6. Alquist, R.; Kilian, L.; Vigfusson, R.J. Forecasting the price of oil. *Handb. Econ. Forecast.* **2013**, *2*, 427–507.
7. Baumeister, C.; Kilian, L. Forecasting the real price of oil in a changing world: A forecast combination approach. *J. Bus. Econ. Stat.* **2015**, *33*, 338–351. [CrossRef]
8. Zhang, J.L.; Zhang, Y.J.; Zhang, L. A novel hybrid method for crude oil price forecasting. *Energy Econ.* **2015**, *49*, 649–659. [CrossRef]
9. Xie, W.; Yu, L.; Xu, S.; Wang, S. A New Method for Crude Oil Price Forecasting Based on Support Vector Machines. *Comput. Sci. ICCS* **2006**, *3994*, 444–451.
10. Yu, L.; Wang, S.; Lai, K.K. Forecasting crude oil price with and EMD-based neural network ensemble learning paradigm. *Energy Econ.* **2008**, *30*, 2623–2635. [CrossRef]
11. Jammazi, R.; Aloui, C. Crude oil price forecasting: Experimental evidence from wavelet decomposition and neural network modeling. *Energy Econ.* **2012**, *34*, 828–841. [CrossRef]
12. Gabralla, L.A.; Jammazi, R.; Abraham, A. Oil price prediction using ensemble machine learning. In Proceedings of the 2013 International Conference on Computing, Electrical and Electronic Engineering (ICCEEE), Khartoum, Sudan, 26–28 August 2013; pp. 674–679.
13. Gumus, M.; Kiran, M.S. Crude oil price forecasting using XGBoost. In Proceedings of the 2017 International Conference on Computer Science and Engineering (UBMK), Antalya, Turkey, 5–7 October 2017; pp. 1100–1103.
14. Chen, Y.; He, K.; Tso, G.K. Forecasting crude oil prices: A deep learning based model. *Procedia Comput. Sci.* **2017**, *122*, 300–307. [CrossRef]
15. Bussiere, M.; Fratzscher, M. Towards a new early warning system of financial crises. *J. Int. Money Financ.* **2006**, *25*, 953–973. [CrossRef]
16. Ye, M.; Zyren, J.; Shore, J. Forecasting short-run crude oil price using high- and low-inventory variables. *Energy Policy* **2006**, *34*, 2736–2743. [CrossRef]
17. Moshiri, S.; Foroutan, F. Forecasting nonlinear crude oil futures prices. *Energy J.* **2006**, *27*, 81–95. [CrossRef]
18. Guo, X.; Li, D.; Zhang, A. Improved support vector machine oil price forecast model based on genetic algorithm optimization parameters. *AASRI Procedia* **2012**, *1*, 525–530. [CrossRef]
19. Niemira, M.P.; Saaty, T.L. An Analytic Network Process model for financial crisis forecasting. *Int. J. Forecast.* **2004**, *20*, 573–587. [CrossRef]
20. Yu, L.; Lai, K.K.; Wang, S.Y. Currency crisis forecasting with general regression neural networks. *Int. J. Inf. Technol. Decis Mak.* **2006**, *5*, 437–454. [CrossRef]
21. Davis, E.P.; Karim, D. Comparing early warning systems for banking crises. *J. Financ. Stabil.* **2008**, *4*, 89–120. [CrossRef]
22. Babecky, J.; Havranek, T.; Mateju, J.; Rusnak, M.; Smidkova, K.; Vasicek, B. Banking, debt and currency crises: Early warning indicators for developed countries. *ECB Work. Pap.* **2012**, 2162901. Available online: https://ssrn.com/abstract=2162901 (accessed on 10 February 2020). [CrossRef]
23. Dabrowski, J.J.; Beyers, C.; de Villiers, J.P. Systemic banking crisis early warning systems using dynamic Bayesian networks. *Expert Syst. Appl.* **2016**, *62*, 225–242. [CrossRef]
24. Sevim, C.; Oztekin, A.; Bali, O.; Gumus, S.; Guresen, E. Developing an early warning system to predict currency crises. *Eur. J. Oper. Res.* **2014**, *237*, 1095–1104. [CrossRef]

25. Geng, R.; Bose, I.; Chen, X. Prediction of financial distress: An empirical study of listed Chinese companies using data mining. *Eur. J. Oper. Res.* **2015**, *241*, 236–247. [CrossRef]
26. Breiman, L. Random forests. *Mach. Learn* **2001**, *45*, 5–32. [CrossRef]
27. Ohlson, J. Financial ratios and the probabilistic prediction of bankruptcy. *J. Account. Res.* **1980**, *18*, 109–131. [CrossRef]
28. Vapnik, V. *Statistical Learning Theory*; Springer: Berlin/Heidelberg, Germany, 1998.
29. Diaz-Uriarte, R.; de Andres, S.A. Gene selection and classification of microarray data using random forest. *BMC Bioinform.* **2006**, *7*, 3. [CrossRef]
30. Chen, X.W.; Liu, M. Prediction of protein-protein interactions using random decision forest framework. *Bioinformatics* **2005**, *21*, 4394–4400. [CrossRef]
31. Malekipirbazari, M.; Aksakalli, V. Risk assessment in social lending via random forests. *Expert Syst. Appl.* **2015**, *42*, 4621–4631. [CrossRef]
32. Bosch, A.; Zisserman, A.; Munoz, X. Image Classification using random forests and ferns. In Proceedings of the IEEE 11th International Conference on Computer Vision, Rio de Janeiro, Brazil, 14–21 October 2007.
33. Maher, J.J.; Sen, T.K. Predicting bond ratings using neural networks: A comparison with logistic regression. *Intell. Syst. Account. Financ. Manag.* **1998**, *6*, 59–72. [CrossRef]
34. Lee, Y.C. Application of support vector machines to corporate credit rating prediction. *Expert Syst. Appl.* **2007**, *33*, 67–74. [CrossRef]
35. Kim, K.J.; Ahn, H. A corporate credit rating model using multi-class support vector machines with an ordinal pairwise partitioning approach. *Comput. Oper. Res.* **2012**, *39*, 1800–1811. [CrossRef]
36. Friedman, H.J.; Hastie, T.; Tibshirani, R. Additive logistic regression: A statistical view of boosting (With discussion and a rejoinder by the authors). *Ann. Stat.* **2000**, *28*, 337–407. [CrossRef]
37. Friedman, H.J. Greedy function approximation: A gradient boosting machine. *Ann. Stat.* **2001**, *29*, 1189–1232. [CrossRef]
38. Altman, N.S. An Introduction to kernel and Nearest–Neighbor Nonparametric Regression. *Am. Stat.* **1992**, *46*, 175–185.
39. Huang, Z.; Chen, H.; Hsu, C.-J.; Chen, W.-H.; Wu, S. Credit rating analysis with support vector machines and neural networks: A market comparative study. *Decis. Support Syst.* **2004**, *37*, 543–558. [CrossRef]
40. Bennell, J.A.; Crabbe, D.; Thomas, S.; Gwilym, O. Modelling sovereign credit ratings: Neural networks versus ordered probit. *Expert Syst. Appl.* **2006**, *30*, 415–425. [CrossRef]
41. Kim, Y.H.; Chang, H.W. Forecasting the volatility of stock price index: A hybrid model integrating LSTM with multiple GARCH-type models. *Expert Syst. Appl.* **2018**, *103*, 25–37. [CrossRef]
42. Zhong, X.; Enke, D. Forecasting daily stock market return udsing dimensionality reduction. *Expert Syst. Appl.* **2007**, *67*, 126–139. [CrossRef]
43. Tanaka, K.; Higashide, T.; Kinkyo, T.; Hamori, S. Analyzing industry-level vulnerability by predicting financial bankruptcy. *Econ Inq.* **2019**, *57*, 2017–2034. [CrossRef]
44. Liaw, A.; Wiener, M. Classification and regression by random forest. *R News* **2002**, *2*, 18–22.
45. Schwarz, D.F.; Konig, I.R.; Ziegler, A. On safari to random jungle: A fast implementation of random forests for high-dimensional data. *Bioinformatics* **2010**, *26*, 1752–1758. [CrossRef] [PubMed]
46. Wright, M.M.; Ziegler, A. ranger: A fast implementation of random forests for high dimensional data in C++ and R. *J. Stat. Softw.* **2015**, *77*, 1–17. [CrossRef]
47. Dong, G.; Lai, K.K.; Yen, J. Credit scorecard on logistic regression with random coefficients. *Procedia Comput. Sci.* **2010**, *1*, 2463–2468. [CrossRef]
48. Cortes, C.; Vapnik, V. Support-vector networks. *Mach. Learn* **1995**, *20*, 273–297. [CrossRef]
49. Bekkar, M.; Kheliouane, H.; Taklit, A. Evaluation measures for models assessment over imbalanced data sets. *J. Inform. Eng. App.* **2013**, *3*, 10.

© 2020 by the authors. Licensee MDPI, Basel, Switzerland. This article is an open access article distributed under the terms and conditions of the Creative Commons Attribution (CC BY) license (http://creativecommons.org/licenses/by/4.0/).

Article

Do Machine Learning Techniques and Dynamic Methods Help Forecast US Natural Gas Crises?

Wenting Zhang and Shigeyuki Hamori *

Graduate School of Economics, Kobe University, 2-1 Rokkodai, Nada-Ku, Kobe 657-8501, Japan; zhangwenting.kobe@gmail.com
* Correspondence: hamori@econ.kobe-u.ac.jp

Received: 24 March 2020; Accepted: 5 May 2020; Published: 9 May 2020

Abstract: Our study combines machine learning techniques and dynamic moving window and expanding window methods to predict crises in the US natural gas market. Specifically, as machine learning models, we employ extreme gradient boosting (XGboost), support vector machines (SVMs), a logistic regression (LogR), random forests (RFs), and neural networks (NNs). The data set used to develop the model covers the period 1994 to 2019 and contains 121 explanatory variables, including those related to crude oil, stock markets, US bond and gold futures, the CBOE Volatility Index (VIX) index, and agriculture futures. To the best of our knowledge, this study is the first to combine machine learning techniques with dynamic approaches to predict US natural gas crises. To improve the model's prediction accuracy, we applied a suite of parameter-tuning methods (e.g., grid-search) to select the best-performing hyperparameters for each model. Our empirical results demonstrated very good prediction accuracy for US natural gas crises when combining the XGboost model with the dynamic moving window method. We believe our findings will be useful to investors wanting to diversify their portfolios, as well as to policymakers wanting to take preemptive action to reduce losses.

Keywords: dynamic approaches; forecasting; logistic regression; random forests; support vector machines; US natural gas crises; XGboost; neural networks

---

1. Introduction

To date, four international energy crises have occurred: The oil crisis of 1973, caused by the Middle East War; the Iranian oil crisis of 1979, caused by the Islamic Revolution; the 1990 oil crisis, caused by the Gulf War; and the energy crisis in 2008 caused by the financial crisis. In addition to being a subsystem of the national economy, the energy system incorporates the petroleum, coal, natural gas, and power economy. Birol [1] showed that the 2008 global financial crisis caused the energy demand in most countries to decrease, ultimately leading to the energy crisis. On the other hand, imbalances between energy supply and demand are still the root cause of rising energy prices, and high energy prices were an underlying cause of the financial crisis.

Compared with fossil fuel oil and coal, natural gas is a more environment friendly and green clean energy. As a result, natural gas has gained an increasingly prominent role in the global energy market. As we know, people always use the settlement prices at the Henry Hub as benchmarks for the entire North American natural gas market and parts of the global liquid natural gas (LNG) market. According to the International Energy Agency (IEA), natural gas had a remarkable year in 2018, with a 4.6% increase in consumption, accounting for nearly half of the increase in global energy demand (https://www.iea.org/fuels-and-technologies/gas). Accurate predictions of crises in the natural gas market make it possible for investors and policymakers to minimize losses. Figure 1 shows some of the major events that affected US natural gas prices between 1994 and 2019.

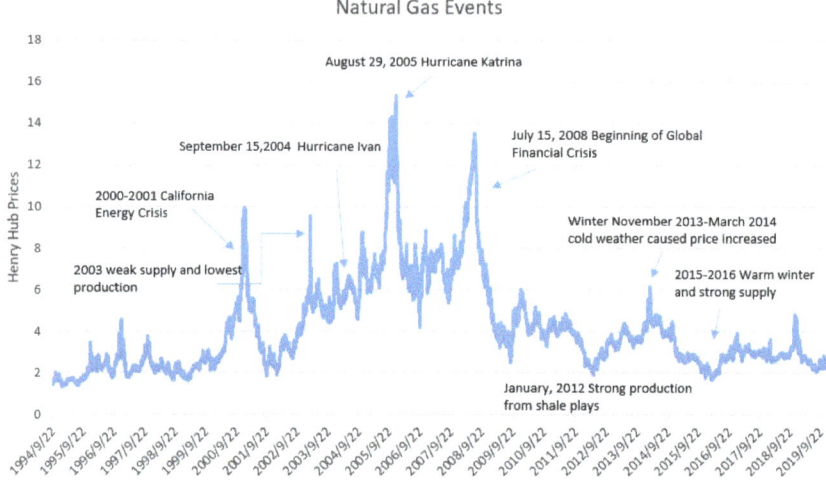

Figure 1. The major crisis events related to US natural gas.

An early warning system (EWS) is a chain of information communication systems comprising sensors, event detection, and decision subsystems. By employing such systems to detect crises before they occur, we can reduce false alarms [2]. An EWS must be able to send a clear signal about whether an economic crisis is impending to complement the expert judgment of decision-makers. Edison [3] developed an operational EWS that successfully detects financial crises based on a signal approach [4,5]. Ideally, an EWS should not ignore any crisis events but should minimize false alarms. However, the cost of not sending a global crisis signal is much higher than the cost of an incorrect alert.

Machine learning refers to a scientific model in which computer systems rely on patterns and inference to perform specific tasks without applying explicit instructions. Machine learning is becoming increasingly useful in finance. As banks and other financial institutions work to strengthen security, streamline processes, and forecast crises, machine learning is becoming their technology of choice. Lin et al. [6] conducted a survey on machine learning in financial crisis prediction, using a machine learning algorithm to investigate the achievements and limitations of cy-prediction and credit-scoring models.

In this study, we propose a dynamic moving window and expanding window methodology that incorporates extreme gradient boosting (XGboost), a support vector machine (SVM), a logistic regression (LogR), a random forest (RF), and a neural networks (NNs) as machine learning methods to predict US natural gas crises. To the best of our knowledge, this study is the first to combine dynamic methodologies with machine learning to predict such crises. The main contributions of this study are as follows:

- We employ dynamic methodologies to define a crisis, thus preventing extreme crisis events.
- We use advanced machine learning techniques, including XGboost, neural networks, and other traditional machine learning methods.
- We combine dynamic methods with machine learning techniques to increase the prediction accuracy of the model.
- We employ novel model evaluation methodologies, and use daily data for the period 1994 to 2019. The long period ensures enough data for the machine learning model, and allows us to check whether US natural gas crises are persistent and clustered.

Our main conclusion is that the combination of XGboost and the dynamic moving window approach performs well in predicting US natural gas crises, particularly in the partial variable case with a moving window. In addition, LogR in the moving window method and NN in the expanding window method do not perform badly in a partial variables situation whereas SVM in the moving window and LogR in the expanding window performance do not perform badly in the all variables situation.

The remainder of the paper proceeds as follows. Section 2 reviews relevant empirical works. In Section 3, we present the data and develop the model. In Section 4, we provide a brief description and technical analysis of the machine learning models employed in this study. In Section 5, we introduce novel model evaluation approaches. Here, we also present the empirical results for each machine learning model. Section 6 concludes the paper. The Appendix A provides the XGboost variable importance plots, the parameter tuning for the postulated SVM plots, the estimated LogR model results, and the RF variable importance plots.

2. Literature Review

Recent financial crises have highlighted their devastating effects on the economy, society, and politics. Therefore, being able to predict crises and disasters, particularly in areas, such as banking, finance, business, medical, and others, means we can implement appropriate measures in advance, and thus minimize losses. Chen et al. [7] considered value-at-risk (VaR) forecasting using a computational Bayesian framework for a portfolio of four Asia-Pacific stock markets, showing that the Autoregressive Conditional Heteroscedasticity (GARCH) model outperforms stochastic volatility models before and after the crises. Bagheri et al. [8] proposed a new hybrid intelligence approach to predict financial periods for the foreign exchange market, finding that it performs well for price forecasting. Niemira and Saaty [9] developed an imbalanced crisis turning point model based on the analytical network process framework to forecast the probability of crises, finding this method to be more flexible and comprehensive than traditional methods. Chiang et al. [10] showed that traders could generate higher returns by employing their proposed adaptive decision support system model.

An energy crisis refers to an energy shortage caused by rising energy prices, and energy and financial crises have mutual effects. Gorucu [11] employed an artificial neural network (ANN) to evaluate and predict natural gas consumption in Ankara, Turkey, examining the factors affecting the output, and then training the ANN to determine the optimal parameters to predict gas consumption, achieving a good performance. Xie et al. [12] used an SVM to forecast crude oil prices, finding that it outperforms both auto regressive integrated moving average and back-propagation neural network models.

Regarding crisis prediction, there are currently three international financial crisis early warning models. First, the discrete choice (probit or logit) model, developed by Frankel and Rose [13], has been used to model financial contagion by Bae et al. [14]. Second, Sachs et al. [15] developed an STV cross-section regression model, finding that the depreciation of the real exchange rate, growth of domestic private loans, and international reserves/M2 (a calculation of the money supply that includes cash and checking deposits as well as near money) are important indicators of whether a country will experience a financial crisis. Third, Kaminsky and Reinhart [5] developed a signaling approach-KLR model that monitors the evolution of several indicators that approach to show unusual behavior in periods preceding a crisis. Furthermore, Knedlik and Schweinitz [16] investigated debt crises in Europe by applying a signal approach, finding that a broad composite indicator has the highest predictive power. Numerous works have since extended these three models, thus improving the early warning system. Based on the multinomial logit model, Bussiere and Fratzscher [17] developed a new early warning system (EWS) model for financial crises. Their results show that employing the proposed EWS substantially improves the ability to forecast financial crises, and that their model would have predicted crises correctly in emerging markets between 1993 and 2001. Xu et al. [18] combined an RF and a wavelet transform in a new EWS to forecast currency crises, finding that real exchange rate appreciation and overvaluation can be measured over a period of 16 to 32 months. Saleh et al. [19] applied a new EWS model for systemic financial fragility and near crises in The Organisation for Economic Co-operation and Development (OECD) countries over a 27-year period, finding that the model has significant implications for financial stability and regulation.

Researchers are increasingly combining EWSs with machine learning techniques to predict financial crises. Ahn et al. [20] extended the EWS classification method to a traditional-type crisis, using an SVM to forecast financial crises. As such, they proved that an SVM is an efficient classifier. Sevim et al. [21] developed three EWSs, based on an ANN, a decision tree, and a LogR model, respectively, to predict currency crises, finding that the decision tree model can predict crises up to a year ahead with approximately 95% accuracy. In order to predict currency crises, Lin et al. [22] developed a hybrid causal model, which is integrated by the learning ability of a neural network with the inference mechanism of a fuzzy logit model, showing that this approach is promising in terms of preventing currency crises. Finally, Chatzis et al. [23] employed a classification tree, SVM, RF, neural network, XGboost, and deep neural network to forecast stock market crises. They show that deep neural networks increase the classification accuracy significantly, and propose an efficient global systemic early warning tool.

Few studies have used machine learning methods with moving windows to predict financial crises. However, studies have focused on other classification methods. Bolbol et al. [24] combined a moving window method with an SVM to classify Global Positioning System (GPS) data into different transportation modes. Chou and Ngo [25] combined the time-series sliding window with the machine learning system to predict real-time building energy-consumption data, showing that the model can potentially predict building energy consumption using big data.

From the aforementioned literature, we can see that the methodologies proposed here offer clear innovations and advantages. Thus, we compared the predictive ability of the following models: XGboost, a SVM, a LogR, a RF, and a NN. In addition, we used a moving window and an expanding window approach. Furthermore, we investigated the effects of explanatory variables on the models' predictive ability by splitting the variables into two categories—natural gas-related predictors (16 predictors), and all predictors (121 predictors)—to compare their prediction accuracy separately. To the best of our knowledge, our study is the first to analyze US natural gas market crises by combining machine learning techniques with dynamic forecasting methods (the moving window and expanding window method).

3. Data

3.1. Data Collection

We employed daily data to model US natural gas market crises, focusing on the US crude oil, stock, commodity, and agriculture markets, where the transmission of extreme events is stronger. In order to obtain an adequate daily data sample for machine learning, we collected data from many different sources and databases (see Table 1).

Table 1. Variables in the model.

Variables	Raw Data	Data Source
Natural Gas	Henry Hub Natural Gas Futures (USD/Million Btu)	Bloomberg
Crude oil	Crude Oil WTI Futures (USD/Barrel)	Bloomberg
Stock	S&P 500 Index	Bloomberg
Bond	United States 10-Year Bond Yield	Bloomberg
Gold	Gold Futures Price	Bloomberg
Vix	CBOE Volatility Index	Bloomberg
Canola	Canola Futures Price	Bloomberg
Cocoa	US Cocoa Futures Price	Bloomberg
Sugar	No.11 Sugar Futures Price	Bloomberg
Rice	Rough Rice Futures Price	Bloomberg
Corn	Corn Futures Price	DataStream
Wheat	US Wheat Futures Price	DataStream

The raw data included the most sensitive market variables from the energy, stock, commodity, and agriculture markets. To remove the effect of the exchange rate on our results, we consolidated the currency units of the variables into a dollar currency unit. Then, because natural gas has the shortest period of daily data (8 June 1990–18 December 2019), we focused on this period for all variables. Finally, we obtained a raw daily price data set with 7152 records. This data set enabled us to model contagion dynamics and financial market interdependence across natural gas-related variables and over time. In order to facilitate the reading of this paper, we named the professional items as exhibited in Table 2.

Table 2. Nomenclature for professional items.

Nomenclature	
p_t	Natural Gas Daily Price
r_t	Natural Gas Daily Return
μ_t	1000-Day Moving and Expanding Window Average return at time t
σ_t	1000-Day Moving and Expanding Window Standard deviation at time t
NC_t	US Natural gas crises at time t
XGboost	Extreme Gradient Boosting
SVM	Support Vector Machines
LogR	Logistic Regression
RF	Random Forests
NN	Neural Networks
G-mean	The geometric mean
LP	The positive likelihood ratio
LN	The negative likelihood ratio
DP	The discriminant power
BA	Balanced accuracy
WBA	The weighted balance accuracy
DM-test	Diebold-Mariano Test

3.2. Data Processing

In Section 3.1, we described the raw data and the prices of the variables. Our goal was to predict next-day crisis events in the US natural gas market using machine learning. Therefore, we needed to define a "crisis event" for natural gas. We also wanted to apply the moving window and expanding window methods in our model. We illustrate these two methods below (see Table 3) [26].

Table 3. Daily data set for the moving window and expanding window.

Variable Names	Numerical Formula
Natural Gas Daily Price	p_t
Natural Gas Daily Return	$r_t = \ln(p_t) - \ln(p_{t-1})$
1000-Day Moving Window Average	$\mu_{t+1000} = \frac{\sum_t^{t+999} r_t}{1000}$ $(t = 1, 2, \ldots, 6151)$
1000-Day Expanding Window Average	$\mu_{i+1001} = \frac{\sum_1^{i+1000} r_t}{i+1000}$ $(i = 0, 1, \ldots 6150, t = 1, 2 \ldots 6151)$
1000-Day Moving Window Standard Deviation	$\sigma_{t+1000} = \sqrt{var(\sum_t^{t+999} r_t)}$ $(t = 1, 2, \ldots, 6151)$
1000-Day Expanding Window Standard Deviation	$\sigma_{i+1001} = \sqrt{var(\sum_1^{i+1000} r_t)}$ $(i = 0, 1, \ldots 6150, t = 1, 2 \ldots 6151)$

We set a 1000-day window, covering five years of daily data. First, we employed the logarithmic difference method to compute the daily return of natural gas. Then, in the daily data set, we calculated the initial empirical distribution of the return on 22 September 1994, based on the daily returns of the first 1000 observations (covering the period 11 June 1990–21 September 1994, which is a 1000-day window). Next, for the moving window method, for each subsequent record, we fixed the number of days (1000 days in the daily data set), and moved the window forward by one day. Then, we recalculated the empirical distribution of the returns based on the new period. Thus, for the last observation in the daily data (18 December 2019), the empirical distribution of the returns was based on the period 23 November 2015 to 17 December 2019 (i.e., 1000 days). In contrast, for the expanding window method, for each subsequent record, we increased the size of the window by one day. Then, we again recalculated the empirical distribution of the returns, including the new observation. Hence, for the last observation in the daily data (18 December 2019), the empirical distribution of the returns was based on the period 11 June 1990, to 17 December 2019 (i.e., 7150 days).

Thus, we obtained the preliminary processed data from which to calculate the "crisis events." Bussiere and Fratzscher [17] developed an EWS model based on a multinomial logit model to forecast financial crises, using $EMP_{i,t}$ to define a currency crisis for each country i and period t. Patel and Sarkar [27], Coudert and Gex [28], and Li et al. [29] used $CMAX_t$, defined as the ratio of an index level at month t to the maximum index level for the period up to month t. Here, we employed the return of natural gas, defined as follows:

$$r_t = ln(p_t) - ln(p_{t-1}), \quad (1)$$

where p_t is the price of natural gas on day t. Next, as in Bussiere and Fratzscher [17], we defined NC_t as a crisis event when the natural gas return r_t is less than two standard deviations below its mean value:

$$NC_t = \begin{cases} 1, & if\ NC_t \leq \mu_t - 2\sigma_t \\ 0, & otherwise \end{cases}, \quad (2)$$

where μ_t and σ_t were calculated based on Table 3. The former is the mean value calculated by the moving window or expanding window on day t. Similarly, the latter denotes the standard deviation calculated by the corresponding methods. We plotted the crises processing procedure, as shown in Figure 2. After processing these data, we obtained two predicted/dependent variables in the daily data set. These constitute binary indicators that take the value of 1 when a crisis occurs in the US natural gas market on the following day, and 0 otherwise. Based on these formulae, Table 4 and Figure 3 summarize the crisis events between 22 September 1994 and 18 December 2019.

Table 4. Crisis events in the US natural gas market.

Year	Number of Crises	
	Moving Window	Expanding Window
1994	2	2
1995	6	6
1996	16	18
1997	6	8
1998	3	4
1999	2	3
2000	6	7
2001	13	14
2002	3	3
2003	8	8
2004	3	5
2005	1	1
2006	7	8
2007	2	2
2008	7	5
2009	9	9
2010	0	0
2011	1	0
2012	3	2
2013	1	0
2014	10	4
2015	6	2
2016	7	0
2017	4	2
2018	10	7
2019	4	3
Total	140	123

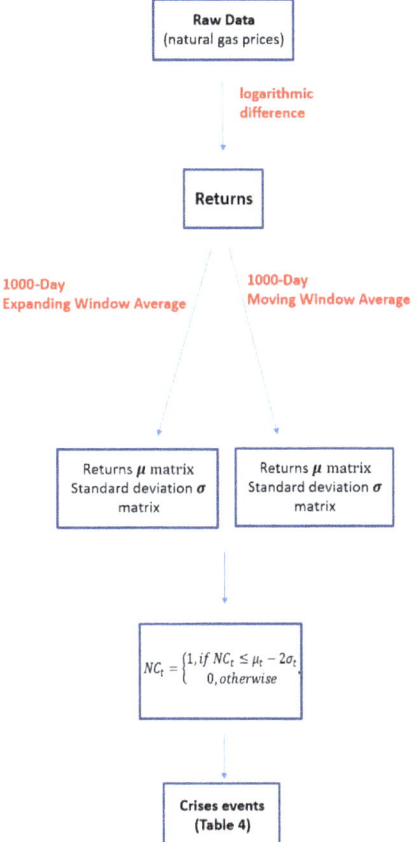

Figure 2. US natural gas crises processing.

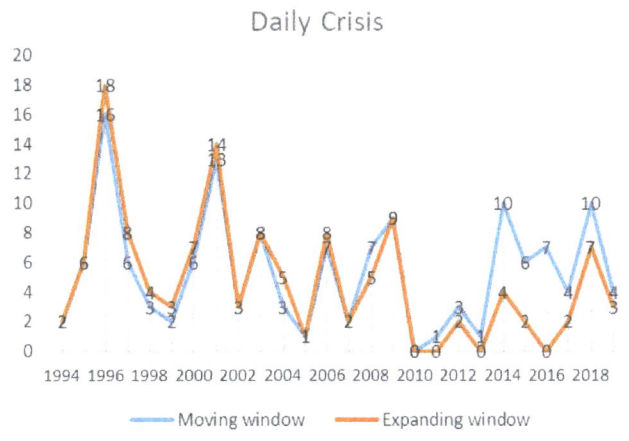

Figure 3. Number of crises in the US natural gas market.

We have outlined the exploratory binary variables in the daily data set using the moving window and expanding window methods. Next, we constructed the independent variables for our machine learning model. In particular, to capture the subtle dynamics and correlations, we performed the following transformation:

1. We calculated the average number of crises during the previous five working days, based on the total number of crisis events.
2. We used the logarithmic difference method to derive the continuously compounded daily return for the prices of natural gas, crude oil, bond yield, gold, canola, cocoa, sugar, rice, corn, and wheat.
3. We calculated the lagged variables for each crisis indicator, for one to five days. Specifically, we considered these lags (lag1 to lag5) for the crisis itself, the returns on all variables except the VIX index, and the returns on all variables including the VIX index.

After the above processing, we obtained two summary tables of daily data for the period 28 February 1994 to 18 December 2019, yielding 6146 records and 121 predictors. The first table determines crises using the moving window method, and the second uses the expanding window method.

4. Model Development

As already discussed, we combined the moving window method, expanding window method, and machine learning to predict crisis events. In addition, we used a moving window and expanding window in the data processing step. Therefore, we employed corresponding methods to develop our model using machine learning. By doing so, we derived dynamic results and improved the accuracy of the model prediction [25]. To maintain the initial number of windows, we employed 1000 days as the initial number of days for the moving window (fixed) and the expanding window (increasing by one day). In order to combine machine learning with the two window methods, we inserted loop statements into the machine learning model to perform 5146 iterations. For both methods, we set all data in the window as the training data set, and set the day following the final day in the training data set as the test data set. For example, in the final loop calculation, the training data set for the moving window method covers the period 19 November 2015 to 17 December 2019 (1000 days), and the test data is that of 18 December 2019; for the expanding window method, the training data set covers the period 29 September 1994 to 17 December 2019 (6145 days), and the test data is that of 18 December 2019. Therefore, we derived a confusion matrix for crisis prediction once the loops were complete.

Next, we chose our methods following Chatzis et al. [23], deciding on the following: XGboost [30], SVM [31], LogR [32], RF [33–35], and NN [36]. We split the explanatory variables into two types to compare the predictions: A natural gas-related variable (the average number of crises during the previous five working days, lags 1 to 5 of crises, and lags 1 to 5 of the prices and returns of natural gas, yielding 16 predictors), and all variables (121 predictors, including natural gas-related variables). In the following section, we describe the development process and the parameter tuning for each model.

4.1. Extreme Gradient Boosting

XGboost is not only an improvement algorithm on the boosting algorithm based on gradient-boosted decision trees, and the internal decision tree uses a regression tree, but also an efficient and prolongable development of the gradient boosting framework [37]. It supports multiple kinds of objective functions, like regression, classification, ranking, and so on. Furthermore, in terms of model accuracy, efficiency, and scalability, XGboost outperforms RFs and NNs [23]. In our study, we employed the XGboost R package developed by Chen and He [30].

In order to increase the accuracy of the model prediction, we employed a (five-fold) cross-validation method to debug a series of vital hyperparameters. In order to build the classification tree, we determined the maximum depth of the tree, minimum number of observations in a terminal node, and size of the subsampling for both window methods. We also tuned the hyperparameter γ, which controls the minimum reduction in the loss function required to grow a new node in a tree, and reduces overfitting the model. In addition, we tuned α (the L1 regulation term on the weights) and λ (the L2 regulation term on weights). Because of the binary nature of our dependent variable, we applied the LogR objective function for the binary classification to train the model. We also calculated the area under the curve (AUC) of our model, which is equal to the probability that the classifier ranks randomly selected positive instances higher than randomly selected negative instances. In our model training, when we employed the moving window method, the AUC values were above 0.8, regardless of whether we used 16 natural gas-related explanatory variables or 121 explanatory variables, indicating a very good performance. However, we derived an AUC of only 0.67 when applying the expanding window method, indicating a poor performance for the combination of this and the XGboost method.

Because we used the moving window and expanding window methods in our study, we cannot show every XGboost variable importance plot, because there were more than 5000 plots after the loop calculation. Therefore, we present the plots for the last step of the loop in the Appendix A. Here, Figures A1 and A2 show that lag5 of the natural gas price is the most important variable in the moving window method. However, lag2 of the natural gas return is the most important in the expanding window method.

4.2. Support Vector Machines

Erdogan [38] applied an SVM [31] to analyze the financial ratio in bankruptcy, finding that the SVM with a Gaussian kernel is capable of predicting bankruptcy crises. This is because the SVM has a regularization parameter, which controls overfitting. Furthermore, using an appropriate kernel function, we can solve many complex problems. An SVM scales relatively well to high-dimensional data. Therefore, it is widely used in financial crisis prediction, handwriting recognition, and so on.

In our study, we used sigmoid, polynomial, radial basis function, and linear kernels to evaluate a soft-margin SVM, finding that the sigmoid function performs optimally. In order to select the optimal hyperparameter (e.g., γ, the free parameter of the Gaussian radial basis function, and c, the cost hyperparameter of the SVM, which involves a trading error penalty for stability), we applied cross-validation. We selected the optimal values of these hyperparameters using the grid-search methodology. We implemented the SVM model using the e1071 package in R, employing the tuning function in the e1071 package to tune the grid-search hyperparameter.

Unlike the XGboost methodology, we could tune the parameters γ and c of the daily data set before executing the loop calculation. The plots of the parameter tuning for the SVM model are provided in the Appendix A. Figures A3 and A4 show the hyperparameters γ (x-axis) and c (y-axis) using the grid-search method in the SVM model. The parameter c is the cost of a misclassification. As shown, a large c yields a low bias and a high variance, because it penalizes the cost of a misclassification. In contrast, a small c yields a high bias and a low variance. The goal is to find a balance between "not too strict" and "not too loose". For γ, the parameter of the Gaussian kernel, low values mean "far" and high values mean "close". When γ is too small, the model will be constrained and will not be able to determine the "shape" of the data. In contrast, an excessively large γ will lead to the radius of the area of influence of the support vectors only including the support vector itself. In general, for an SVM, a high value of gamma leads to greater accuracy but biased results, and vice versa. Similarly, a large value of the cost parameter (c) indicates poor accuracy but low bias, and vice versa.

4.3. Logistic Regression

In statistics, the LogR algorithm is widely applied for modeling the probability of a certain class or event existing, such as crisis/no crisis, healthy/sick or win/lose, and so on. The LogR is a statistical model, which applies a logistic function to model a binary dependent variable in its basic form. Ohlson [32] employed a LogR to predict corporate bankruptcy using publicly available financial data. Yap et al. [39] employed a LogR to predict corporate failures in Malaysia over a 10-year period, finding the method to be very effective and reliable. In our study, we employed the glm function in R, following

Hothorn and Everitt [40]. In addition, because the output of a LogR is a probability, we selected optimistic thresholds for the two window methods separately when predicting crises. We also employed a stepwise selection method to identify the statistically significant variables. However, this did not result in a good performance. Therefore, we did not consider the stepwise method further.

As before, we summarize some of the outcomes of our analysis in the Appendix A. Here, Tables A1 and A2 exhibit some of the outcomes of our analysis and we also present the plot of the final step of the loop calculation, showing only the intentionality of some of the variables. As shown, whether in the partial variables case or the all variables case, the lagged value and the average of five days of crisis are the most intentional.

4.4. Random Forests

An RF [33–35] is an advanced learning algorithm employed for both classification and regression created by Ho [41]. A forest in the RF model is made up of trees, where a greater number of trees denotes a more robust forest. The RF algorithm creates decision trees from data samples, obtains a prediction from each, and selects the best solution by means of voting. Compared with a single decision tree, this is an ensemble method. Because it can reduce overfitting by averaging the results. RFs are known as bagging or bootstrap aggregation methods. By identifying the best segmentation feature from the random subset of available features, further randomness can be introduced.

In RF, a large number of predictors can provide strong generalization; our model contains 121 predictors. In order to improve the accuracy of model prediction, we must obtain an optimal value of m, the number of variables available for splitting at each tree node. If m is relatively low, then both the inter-tree correlation and the strength of each tree will decrease. To find the optimal value, we employed the grid-search method. In addition to m, we tuned the number of trees using the grid-search method. To reduce the computing time, we applied the ranger package in R to construct the prediction model, employing the tuneRF function in the randomForest package.

We provide the RF variable importance plots in the last loop calculation in Appendix A. In Figures A5 and A6, we present the importance of each indicator for the classification outcome. The obtained ranking is based on the mean decrease in the Gini. This is the average value of the total reduction of variables in node impurities, weighted by the proportion of samples that reach the node in each decision tree in RF. Here, a higher mean decrease in the Gini indicates higher variable importance. We found that in the partial variables case, the lags of the natural gas price and the natural gas return are relatively important; however, in the case of all variables, this was not the case.

4.5. Neural Networks

In the field of machine learning and cognitive science, an NN [34] is a mathematical or computational model, which imitates the structure and function of a biological neural network (animal central nervous system, especially the brain), employed to estimate or approximate functions. Like other machine learning methods, NN have been used to solve a variety of problems, such as credit rating classification problems, speech recognition, and so on. The most usually considered multilayer feedforward network is composed of three types of layers. First, in the input layer, many neurons accept a large number of nonlinear input messages. The input message is called the input vector. Second, the hidden layer, which is the various layers composed of many neurons and links between the input layer and output layer. The hidden layer can have one or more layers. The number of neurons in the hidden layer is indefinite, but the greater the number, the more nonlinear the NN becomes, so that the robustness of the neural network is more significant. Third, in the output layer, messages are transmitted, analyzed, and weighed in the neuron link to form the output. The output message is called the output vector.

In order to increase the accuracy of the model prediction, we employed the grid-search method to find the best parameter size and decay. The parameter size is the number of units in the hidden layer and parameter decay is the regularization parameter to avoid over-fitting. We used nnet package in R to develop the model prediction.

5. Model Evaluation

In this section, we evaluate the robustness of our methods by performing a comprehensive experimental evaluation procedure. Specifically, we use several criteria to evaluate the aforementioned models in terms of crisis event prediction.

5.1. Verification Methods

Accuracy is measured by the discriminating power of the rating power of the rating system. It is the most commonly used indicator for classification, and it estimates the overall effectiveness of an approach by estimating the probability of the true value of the class label. In the following section, we introduce a series of indicators that are widely employed to quantitatively estimate the discriminatory power of each model.

In machine learning, we can derive a square 2 × 2 matrix, as shown in Table 5, which we used as the basis of our research.

Table 5. Confusion matrix for two classes classification.

	Predicted 0	Predicted 1
Actual 0	TN (number of True Negatives)	FP (number of False Positives)
Actual 1	FN (number of False Negatives)	TP (number of True Positives)

Notes: 0 refers to "No crisis"; 1 refers to "Crisis"; Predicted 0 refers to a predicted value of 0, Predicted 1 refers to a predicted value of 1; Actual 0 refers to a true value of 0; Actual 1 refers to a true value of 1; TN = True Negative: the number of negative cases (i.e., no crisis) correctly identified as negative cases; FP = False Positive: the number of negative cases (i.e., no crisis) incorrectly identified as positive cases; FN = False Negative: the number of positive cases (i.e., crisis) misclassified as negative cases; TP = True Positive: the number of positive cases (i.e., crisis) correctly identified as positive.

Bekkar et al. [42] state that imbalanced data learning is one of the challenging problems in data mining, and consider that skewed classes in the level distribution can lead to misleading assessment methods and, thus, bias in the classification. To resolve this problem, they present a series of alternatives for imbalanced data learning assessment, which we introduce below.

Bekkar et al. [42] defined sensitivity and specificity as follows:

$$Sensitivity\ (or\ Recall) = \frac{TP}{TP+FN};\ Specificity = \frac{TN}{TN+FP}. \tag{3}$$

Sensitivity and specificity assess the effectiveness of an algorithm on a single class for positive and negative outcomes, respectively. The evaluation measures are as follows:

- G-mean: The geometric mean is the product of the sensitivity and specificity. This metric can indicate the valance between the classification performance of the majority and minority classes. In addition, the G-mean measures the degree to which we avoid overfitting on the negative class, and the degree to which the positive class is marginalized. The formula is as follows:

$$G-mean = \sqrt{Sensitivity \times Specificity}. \tag{4}$$

Here, a low G-mean value means the prediction performance of the positive cases is poor, even if the negative cases are correctly classified by the algorithm.

- LP: The positive likelihood ratio is that between the probability of predicting an example as positive when it is really positive, and the probability of a positive prediction when it is not positive. The formula is as follows:

$$LP = \frac{TP/(TP+FN)}{FP/(FP+TN)} = \frac{Sensitivity}{1-Specificity}. \tag{5}$$

- LN: The negative likelihood ratio is that between the probability of predicting an example as negative when it is really positive, and the probability of predicting a case as negative when it is actually negative. The formula is as follows:

$$LN = \frac{FN/(TP+FN)}{TN/(FP+TN)} = \frac{1-Sensitivity}{Specificity}. \tag{6}$$

- DP: The discriminant power, a measure that summarizes sensitivity and specificity. The formula is as follows:

$$DP = \frac{\sqrt{3}}{\pi}\left(\log\frac{Sensitivity}{1-Sensitivity} - \log\frac{Specificity}{1-Specificity}\right). \tag{7}$$

DP values can evaluate the degree to which the algorithm differentiates between positive and negative cases. DP values lower than one indicate that the algorithm differentiates poorly between the two, and values higher than three indicate that it performs well.

- BA: Balanced accuracy, calculated as the average of the sensitivity and specificity. The algorithm is as follows:

$$BA = \frac{1}{2}(Sensitivity + Specificity). \tag{8}$$

If the classifier performs equally well in both classes, this term reduces to the conventional accuracy measure. However, if the classification performs well in the majority class (in our study, "no crisis"), the balance accuracy will drop sharply. Therefore, the BA considers the majority class and minority class (in our study, "crisis") equally.

- WBA: The weighted balance accuracy. Under the weighting scheme 75%: 25%, the WBA emphasizes sensitivity over specificity, and is defined by Chatzis et al. [23] as follows:

$$WBA = 0.75 * Sensitivity + 0.25 * Specificity. \tag{9}$$

- Youden's γ: An index that evaluates the ability of an algorithm to avoid failure. This index incorporates the connection between sensitivity and specificity, as well as a linear correspondence with balanced accuracy:

$$\gamma = Sensitivity - (1 - Specificity) = 2 * BA - 1. \tag{10}$$

- F-measure: The F-measure employs the same contingency matrix as that of relative usefulness. Powers [43] shows that an optimal prediction implies an F-measure of one. The formula is as follows:

$$F-measure = \frac{TP}{TP + \frac{(FP+FN)}{2}}. \tag{11}$$

- DM-test: The Diebold–Mariano test is a quantitative approach to evaluate the forecast accuracy of US natural gas crises predicting models in our paper [44]. In our paper, the DM-test can help us to discriminate the significant differences of the predicting accuracy between the XGboost,

SVM, LogR, RF, and NN models based on the scheme of quantitative analysis and about the loss function. We employed the squared-error loss function to do the DM-test.

To simplify the explanation of its values, the F-measure was defined such that a higher value of γ indicates a better ability to avoid classifying incorrectly.

After a series of illustrations of the algorithm above, we derived a variety of methods to evaluate the model prediction ability. Here, we used all of the methods to derive a comprehensive and accurate view of the models' performance. Lastly, we calculated an optimal US natural gas market collapse probability critical point for each fitted model, thus obtaining the optimal sensitivity and specificity measures. We focused on false negatives rather than false positives, because the goal of our work was to construct a supervised warning mechanism, which can predict as many correct signals as possible while reducing the incidence of false negatives.

5.2. Models Prediction Results

We identified 121 predictors and 6146 records for the period 29 September 1994 to 18 December 2019. In addition to using the two window methods, we determined the performance of the models when predicting US natural gas crises using natural gas-related predictors only (16 predictors) and all predictors (121 predictors) separately. In the following, we refer to the combination of the moving window method and the natural gas-related predictors as the "partial variables with the moving window" method. Similarly, we have the "partial variables with expanding window" method. When we combined the moving window method with all variables, we have the "all variables with a moving window" method and, similarly, we have the "all variables with an expanding window" method.

As shown in Tables 6 and 7, when we applied the partial explanatory variables (16 variables) to develop the machine learning models, XGboost provided the best empirical performance for both window methods, followed by the LogR in the moving window method and NN in the expanding window method since the Youden's γ of XGboost was the highest in both two window methods, which means that XGboost has a better ability to avoid misclassification. Furthermore, the BA and WBA of XGboost was also the highest, which means that XGboost performs well in both crises' prediction and no-crisis forecasting. When we employed all explanatory variables (121 variables), XGboost clearly still performed best in both dynamic methods, followed by the SVM in the moving window and LogR in the expanding window from Youden's γ values. Similarly, the BA and WBA of XGboost was also the highest in the all variables situation.

Table 6. Verification methods: Dependent variable concerns a US natural gas crisis (partial variables).

	Partial Variables									
	Moving Window Method					Expanding Window Method				
	XGboost	SVM	LogR	RF	NN		XGboost	SVM	LogR	RF	NN
G-mean	0.617	0.353	0.460	0.328	0.377	G-mean	0.589	0.419	0.493	0.300	0.489
LP	2.523	4.556	3.032	5.384	4.427	LP	2.912	4.135	3.215	3.877	4.618
LN	0.653	0.897	0.834	0.908	0.882	LN	0.694	0.854	0.802	0.930	0.790
DP	0.745	0.896	0.712	0.981	0.889	DP	0.791	0.870	0.766	0.787	0.973
BA	**0.641**	0.550	0.577	0.545	0.557	BA	**0.632**	0.570	0.591	0.534	0.599
WBA	**0.555**	0.339	0.403	0.327	0.352	WBA	**0.517**	0.377	0.428	0.313	0.426
Youden	**0.282**	0.100	0.154	0.090	0.114	Youden	**0.264**	0.139	0.182	0.068	**0.198**
F-measure	0.093	0.106	0.097	0.107	0.110	F-measure	0.085	0.098	0.087	0.074	0.114

Notes: "Partial variables" refers to 16 natural gas-related explanatory variables. G-mean (geometric mean), LP (positive likelihood ratio), LN (negative likelihood ratio), DP (discriminant power), BA (balanced accuracy), weighted balanced accuracy (WBA), Youden (Youden's γ), F-measure.

In addition, because Tables 6 and 7 show that the forecasting performances of XGboost are great and far exceed the other four models, when implementing the DM-test, we only evaluated the forecast accuracy between XGboost and the other four models. From Tables 8 and 9, we can see from the results of DM-test that in both the partial variables and all variables situations and whether by the moving window method or expanding window method, according to the DM-test based on the squared-error loss, all the DM-test values >1.96, the zero hypothesis is rejected at the 5% level of significance; that is to say, the observed differences between XGboost and other four models are significant and XGboost in US natural gas crises prediction accuracy is best between the five models used in our paper.

Table 7. Verification methods: Dependent variable concerns a US natural gas crisis (all variables).

	All Variables										
	Moving Window Method						Expanding Window Method				
	XGboost	SVM	LogR	RF	NN		XGboost	SVM	LogR	RF	NN
G-mean	0.605	0.374	0.387	0.325	0.357	G-mean	0.541	0.460	0.448	0.278	0.330
LP	1.958	2.993	1.797	2.705	1.858	LP	2.742	1.948	2.682	2.035	2.092
LN	0.684	0.897	0.919	0.928	0.931	LN	0.759	0.866	0.851	0.957	0.937
DP	0.580	0.664	0.369	0.590	0.381	DP	0.708	0.447	0.633	0.416	0.443
BA	**0.619**	0.549	0.537	0.535	0.532	BA	**0.606**	0.559	0.568	0.520	0.530
WBA	**0.553**	0.348	0.351	0.322	0.335	WBA	**0.470**	0.400	0.393	0.300	0.322
Youden	**0.238**	0.098	0.073	0.069	0.064	Youden	**0.212**	0.117	**0.137**	0.041	0.060
F-measure	0.075	0.086	0.061	0.074	0.060	F-measure	0.079	0.057	0.073	0.048	0.053

Notes: "All variables" refers to all 121 explanatory variables. G-mean (geometric mean), LP (positive likelihood ratio), LN (negative likelihood ratio), DP (discriminant power), BA (balanced accuracy), weighted balanced accuracy (WBA), Youden (Youden's γ), F-measure.

Table 8. The results of the Diebold–Mariano test (partial variables).

	Partial Variables			
	Moving Window Method			
	XGboost&SVM	XGboost&LogR	XGboost&RF	XGboost&NN
DM-test	30.138	22.655	24.383	25.848
p-value	$p < 2.2 \times 10^{-16}$	$p < 2.2 \times 10^{-16}$	$p < 2.2 \times 10^{-16}$	$p < 2.2 \times 10^{-16}$
	Expanding Window Method			
	XGboost&SVM	XGboost&LogR	XGboost&RF	XGboost&NN
DM-test	20.483	13.871	23.793	21.684
p-value	$p < 2.2 \times 10^{-16}$	$p < 2.2 \times 10^{-16}$	$p < 2.2 \times 10^{-16}$	$p < 2.2 \times 10^{-16}$

Notes: DM-test means Diebold–Mariano test. We employ the squared-error loss function to do the DM-test.

Table 9. The results of the Diebold–Mariano test (all variables).

	All Variables			
	Moving Window Method			
	XGboost&SVM	XGboost&LogR	XGboost&RF	XGboost&NN
DM-test	31.334	20.68	29.858	23.673
p-value	$p < 2.2 \times 10^{-16}$	$p < 2.2 \times 10^{-16}$	$p < 2.2 \times 10^{-16}$	$p < 2.2 \times 10^{-16}$
	Expanding Window Method			
	XGboost&SVM	XGboost&LogR	XGboost&RF	XGboost&NN
DM-test	8.4212	15.986	23.138	20.041
p-value	$p < 2.2 \times 10^{-16}$	$p < 2.2 \times 10^{-16}$	$p < 2.2 \times 10^{-16}$	$p < 2.2 \times 10^{-16}$

Notes: DM-test means Diebold–Mariano test. We employ the squared-error loss function to do the DM-test.

Summarizing the results across the five model evaluations in the test sample, it is evident that XGboost outperforms the other machine learning models. So far, there are very few papers related to forecasting US natural gas crises through machine learning. Thus, our research has made a certain contribution to the prediction of US natural gas crises.

5.3. Confusion Matrix Results

Referring to Chatzis et al. [23], we computed the final classification performance confusion matrices for the evaluated machine learning models (see Tables 10 and 11). Here, a false alarm means a false positive rate, and the hit rate denotes the proportion of correct crisis predictions in the total number of crises. For the best performing model, XGboost, we found that the false alarms do not exceed 25%, and the highest hit rate is 49%. Thus, our US natural gas crisis prediction accuracy can reach 49% using all variables with the moving window method.

Table 10. Confusion matrix results: Partial variables.

Partial Variables									
Moving Window Method					Expanding Window Method				
XGboost	Predict					Predict			
TRUE	0	1	Signal		TRUE	0	1	Signal	
0	4103	934	False alarm	19%	0	4360	699	False alarm	15%
1	58	51	Hit rate	47%	1	52	35	Hit rate	40%
SVM	Predict					Predict			
TRUE	0	1	Signal		TRUE	0	1	Signal	
0	4895	142	False alarm	5%	0	4834	225	False alarm	6%
1	95	14	Hit rate	13%	1	71	16	Hit rate	18%
LogR	Predict					Predict			
TRUE	0	1	Signal		TRUE	0	1	Signal	
0	4656	381	False alarm	9%	0	4643	416	False alarm	9%
1	84	25	Hit rate	23%	1	64	23	Hit rate	26%
RF	Predict					Predict			
TRUE	0	1	Signal		TRUE	0	1	Signal	
0	4934	103	False alarm	4%	0	4939	120	False alarm	4%
1	97	12	Hit rate	11%	1	79	8	Hit rate	9%
NN	Predict					Predict			
TRUE	0	1	Signal		TRUE	0	1	Signal	
0	4870	167	False alarm	5%	0	4782	277	False alarm	7%
1	93	16	Hit rate	15%	1	65	22	Hit rate	25%

Table 11. Confusion matrix results: All variables.

All Variables									
Moving Window Method					Expanding Window Method				
XGboost	Predict					Predict			
TRUE	0	1	Signal		TRUE	0	1	Signal	
0	3786	1251	False alarm	25%	0	4444	615	False alarm	13%
1	56	53	Hit rate	49%	1	58	29	Hit rate	33%
SVM	Predict					Predict			
TRUE	0	1	Signal		TRUE	0	1	Signal	
0	4790	247	False alarm	7%	0	4432	627	False alarm	13%
1	93	16	Hit rate	15%	1	66	21	Hit rate	24%
LogR	Predict					Predict			
TRUE	0	1	Signal		TRUE	0	1	Signal	
0	4574	463	False alarm	11%	0	4647	412	False alarm	9%
1	91	18	Hit rate	17%	1	68	19	Hit rate	22%
RF	Predict					Predict			
TRUE	0	1	Signal		TRUE	0	1	Signal	
0	4832	205	False alarm	6%	0	4859	200	False alarm	5%
1	97	12	Hit rate	11%	1	80	7	Hit rate	8%
NN	Predict					Predict			
TRUE	0	1	Signal		TRUE	0	1	Signal	
0	4664	373	False alarm	9%	0	4781	278	False alarm	7%
1	94	15	Hit rate	14%	1	77	10	Hit rate	11%

6. Conclusions

We proposed a dynamic moving window and expanding window method that combines XGboost, an SVM, a LogR, an RF, and an NN as machine learning techniques to predict US natural gas crises. We proposed the following full procedure. First, we selected the most significant US natural gas financial market indicators, which we believe can be employed to predict US natural gas crises. Second, we combined the aforementioned dynamic methods with other methods (e.g., the returns and lags of the raw prices) to process the daily data set, and defined a crisis for the machine learning model. Third, we used the five machine learning models (XGboost, SVM, LogR, RF, and NN) with dynamic methods to forecast the crisis events. Finally, we evaluated the performance of each model using various validation measures. We demonstrated that combining dynamic methods with machine learning models can achieve a better performance in terms of predicting US natural gas crises. In addition, our empirical results indicated that XGboost with the moving window method achieves a good performance in predicting such crises.

Our main conclusions can be summarized as follows. From the various verification methods, DM-test results, and confusion matrix results, XGboost with a moving window approach achieved a good performance in predicting US natural gas crises, particularly in the partial variables case. In addition, LogR in the moving window method and NN in the expanding window method did not perform badly in the partial variables situation whereas SVM in the moving window and LogR in the expanding window did not perform badly in the all variables situation. Our conclusions can help us to forecast the crises of the US natural gas market more accurately. Because financial markets are contagious, policymakers must consider the potential effect of a third market. This is equally important for asset management investors, because diversified benefits may no longer exist during volatile times. Therefore, these findings can help investors and policymakers detect crises, and thus take preemptive actions to minimize potential losses.

The novel contributions of our study can be summarized as follows. First, to the best of our knowledge, this study is the first to combine dynamic methodologies with machine learning to predict US natural gas crises. A lot of the previous studies are about the currency exchange crises, bank credit crises, and so on. So far, there are very few papers related to forecasting US natural gas crises through machine learning. Second, we implemented XGboost, a popular and advanced machine learning model in the financial crisis prediction field. Additionally, we found that XGboost outperforms in forecasting US natural gas crises. Third, we employed various parameter tuning methods (e.g., the grid-search) to improve the prediction accuracy. Fourth, we used novel evaluation methods appropriate for imbalanced data sets to measure the model performance and we also employed the DM-test method and found that the forecasting performance of XGboost is significant. Finally, we employed in-depth explanatory variables that cover the spectrum of major financial markets.

Before combining the dynamic methodologies with machine learning techniques, we performed ordinary machine learning, in that 70% of the data was used as training data and 30% was used as test data. However, the prediction performance of this method was not so satisfactory. Although our research is the first to combine machine learning techniques with dynamic methodologies for US natural gas crisis prediction, the prediction accuracy of our model can be improved. This is left to future research. Lastly, because economic conditions change continuously and crises continue to occur, predicting crisis events will remain an open research issue.

Author Contributions: Investigation, W.Z.; writing—original draft preparation, W.Z.; writing—review and editing, S.H.; project administration, S.H.; funding acquisition, S.H. All authors have read and agreed to the published version of the manuscript.

Funding: This work was supported by JSPS KAKENHI Grant Number (A) 17H00983.

Acknowledgments: We are grateful to four anonymous referees for their helpful comments and suggestions.

Conflicts of Interest: The authors declare no conflict of interest.

Appendix A

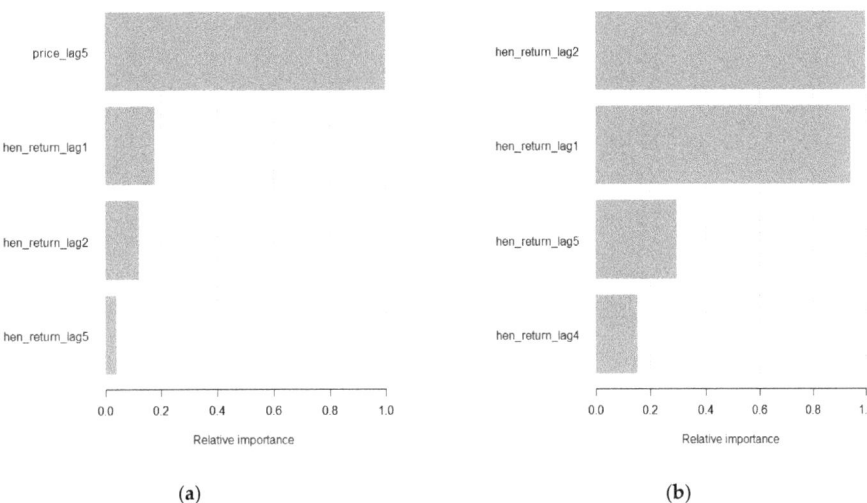

Figure A1. XGboost variable importance plots: partial variables with the moving window and the expanding window methods. (**a**) the XGboost variable importance plot of the partial variables with the moving window method in the final loop calculation; (**b**) the XGboost variable importance plot of the partial variables with the expanding window method in the final loop calculation.

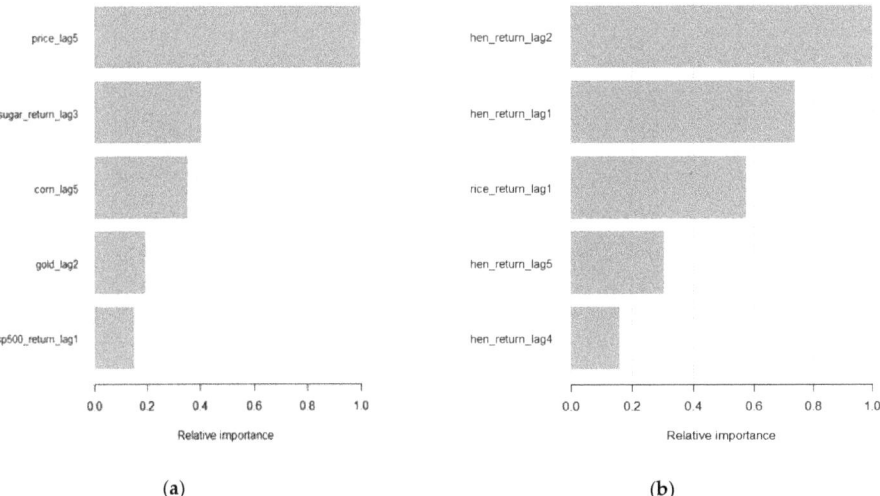

Figure A2. XGboost variable importance plots: all variables with the moving window and the expanding window methods. (**a**) the XGboost variable importance plot of all variables with the moving window method in the final loop calculation; (**b**) the XGboost variable importance plot of all variables with the expanding window method in the final loop calculation.

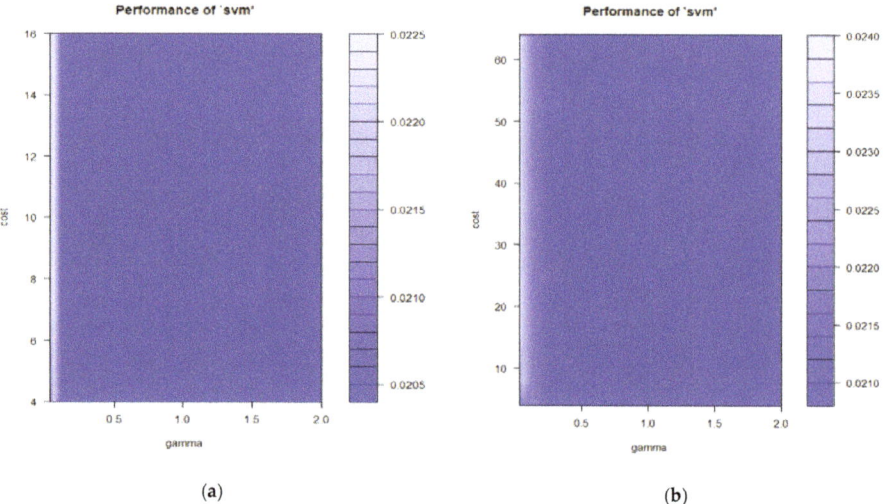

Figure A3. Parameter tuning for the postulated SVM plot: partial variables with the moving window and expanding window methods. (**a**) parameter tuning for the SVM plot of the partial variables with the moving window method in the final loop calculation; (**b**) parameter tuning for the SVM plot of the partial variables with the expanding window method in the final loop calculation.

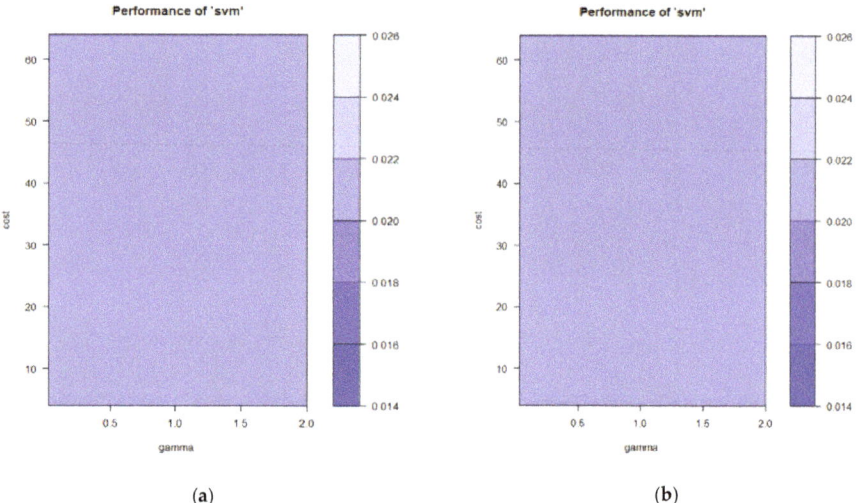

Figure A4. Parameter tuning for the postulated SVM plot: all variables with the moving window and the expanding window methods. (**a**) parameter tuning for the SVM plot of all variables with the moving window method in the final loop calculation; (**b**) parameter tuning for the SVM plot of all variables with the expanding window method in the final loop calculation.

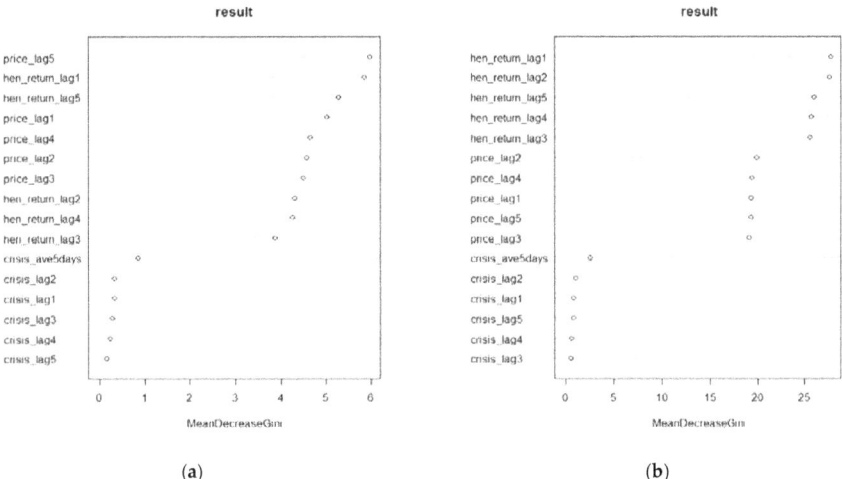

Figure A5. RF variable importance plots: partial variables with the moving window and the expanding window methods. (**a**) RF variable importance plot of the partial variables with the moving window method in the final loop calculation; (**b**) RF variable importance plot of the partial variables with the expanding window method in the final loop calculation.

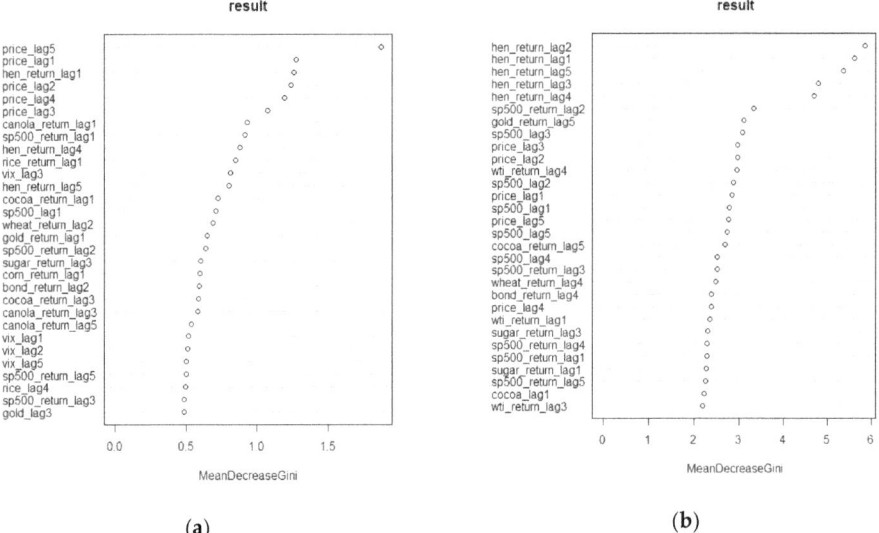

Figure A6. RF variable importance plots: all variables with the moving window and the expanding window methods. (**a**) RF variable importance plot of all variables with the moving window method in the final loop calculation; (**b**) RF variable importance plot of all variables with the expanding window method in the final loop calculation.

Table A1. Estimated LogR model result: Partial variables.

Partial Variables									
Moving Window Method					Expanding Window Method				
	Estimate	Std. Error	z-Value	Pr(>|z|)		Estimate	Std. Error	z-Value	Pr(>|z|)
(Intercept)	−7.91	1.19	−6.65	2.91×10^{-11} ***	(Intercept)	-1.88×10^3	1.12×10^6	−0.002	0.999
crisis_ave5days	2.77	6.50	0.43	0.6698	crisis_ave5days	-8.71×10^2	7.42×10^5	−0.001	0.999
crisis_lag1	2.61	1.56	1.67	0.0942	crisis_lag1	2.19×10^2	1.23×10^5	0.002	0.999
hen_return_lag1	39.25	18.85	2.08	0.0373 *	hen_return_lag1	2.90×10^3	1.31×10^6	0.002	0.998

Notes: *** < 0.001 p-value; * < 0.05 p-value.

Table A2. Estimated LogR model result: All variables.

All Variables									
Moving Window Method					Expanding Window Method				
	Estimate	Std. Error	z-Value	Pr(>|z|)		Estimate	Std. Error	z-Value	Pr(>|z|)
(Intercept)	−4.56	0.21	−21.87	2×10^{-16} ***	(Intercept)	−5.48	1.73	−3.17	0.00153 **
crisis_ave5days	8.75	2.34	3.73	0.000190 ***	crisis_ave5days	7.30	2.58	2.83	0.00465 **
hen_return_lag5	7.31	2.21	3.31	0.000944 ***	hen_return_lag5	6.60	2.39	2.76	0.00582 **
hen_return_lag2	9.19	4.07	2.26	0.023885 *	gold_return_lag5	2.61	9.88	2.64	0.00835 **

Notes: *** < 0.001 p-value; ** < 0.01 p-value; * < 0.05 p-value.

References

1. Birol, F. The Impact of the Financial and Economic Crisis on Global Energy Investments. In Proceedings of the Energy, Policies and Technologies for Sustainable Economies, IAEE (International Association for Energy Economics) European Conference, Vienna, Italy, 7–10 September 2009.
2. Klopotan, I.; Zoroja, J.; Mesko, M. Early warning system in business, finance, and economics: Bibliometric and topic analysis. *Int. J. Eng. Bus. Manag.* **2018**, *10*, 1–12. [CrossRef]
3. Edison, J.H. Do indicators of financial crises work? An evaluation of an early warning system. *Int. J. Financ. Econ.* **2003**, *8*, 11–53. [CrossRef]
4. Kaminsky, L.G.; Lizondo, S.; Reinhart, M.C. Leading indicators of currency crises. *IMF Econ. Rev.* **1998**, *45*, 1–48. [CrossRef]
5. Kaminsky, L.G.; Reinhart, M.C. The twin crises: The causes of banking and balance-of-payments problems. *Am. Econ. Rev.* **1999**, *89*, 473–500. [CrossRef]
6. Lin, W.Y.; Hu, Y.H.; Tsai, C.F. Machine learning in financial crisis prediction: A survey. *IEEE Trans. Syst. Man. Cybern. Part C* **2012**, *42*, 421–436.
7. Chen, S.W.C.; Gerlach, R.; Lin, M.H.E.; Lee, W.C.W. Bayesian forecasting for financial risk management, pre and post the global financial crisis. *J. Forecast.* **2012**, *31*, 661–687. [CrossRef]
8. Bagheri, A.; Peyhani, M.H.; Akbari, M. Financial forecasting using ANFIS networks with quantum-behaved particle swarm optimization. *Expert Syst. Appl.* **2014**, *41*, 6235–6250. [CrossRef]
9. Niemira, P.M.; Saaty, L.T. An analytic network process model for financial-crisis forecasting. *Int. J. Forecast.* **2004**, *20*, 573–587. [CrossRef]
10. Chiang, W.C.; Enke, D.; Wu, T.; Wang, R.Z. An adaptive stock index trading decision support system. *Expert Syst. Appl.* **2016**, *59*, 195–207. [CrossRef]
11. Gorucu, B.F. Artificial neural network modeling for forecasting gas consumption. *Energy Source* **2004**, *26*, 299–307. [CrossRef]
12. Xie, W.; Yu, L.; Xu, S.Y.; Wang, S.Y. A New Method for Crude Oil Price Forecasting Based on Support Vector Machines. In *International Conference on Computational Science*; Springer: Berlin/Heidelberg, Germany, 2006; pp. 444–451.
13. Frankel, A.J.; Rose, K.A. Currency crashes in emerging markets: An empirical treatment. *J. Int. Econ.* **1996**, *41*, 351–366. [CrossRef]
14. Bae, K.H.; Karolyi, A.; Stulz, M.R. A new approach to measuring financial contagion. *Rev. Financ. Stud.* **2003**, *16*, 717–763. [CrossRef]
15. Sachs, J.; Tornell, A.; Velasco, A. Financial crises in emerging markets: The lessons from 1995. *Brook. Pap. Econ. Act.* **1996**, *27*, 147–199. [CrossRef]

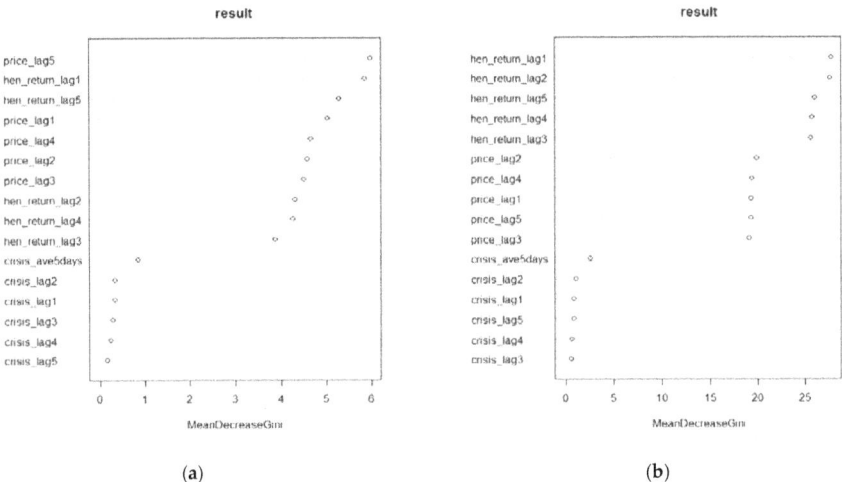

Figure A5. RF variable importance plots: partial variables with the moving window and the expanding window methods. (**a**) RF variable importance plot of the partial variables with the moving window method in the final loop calculation; (**b**) RF variable importance plot of the partial variables with the expanding window method in the final loop calculation.

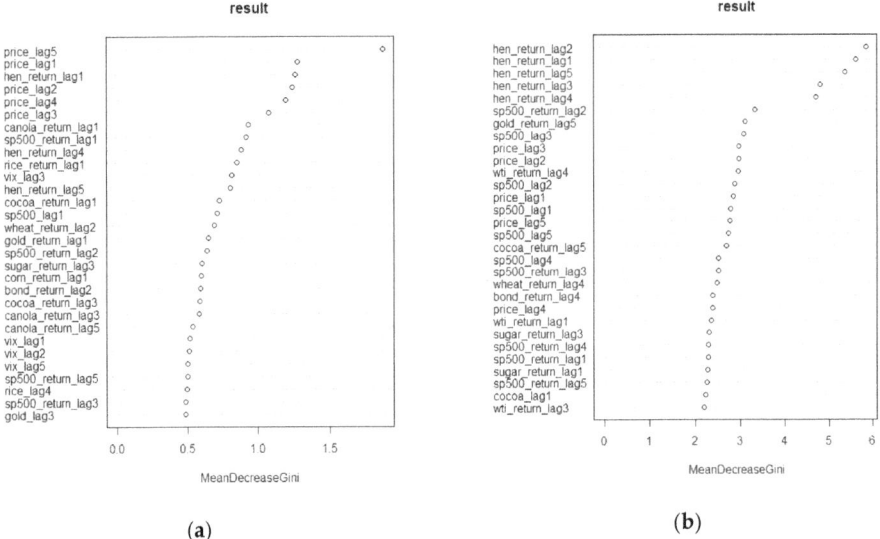

Figure A6. RF variable importance plots: all variables with the moving window and the expanding window methods. (**a**) RF variable importance plot of all variables with the moving window method in the final loop calculation; (**b**) RF variable importance plot of all variables with the expanding window method in the final loop calculation.

Table A1. Estimated LogR model result: Partial variables.

	Partial Variables								
	Moving Window Method					Expanding Window Method			
	Estimate	Std. Error	z-Value	Pr(>|z|)		Estimate	Std. Error	z-Value	Pr(>|z|)
(Intercept)	−7.91	1.19	−6.65	2.91×10^{-11} ***	(Intercept)	-1.88×10^3	1.12×10^6	−0.002	0.999
crisis_ave5days	2.77	6.50	0.43	0.6698	crisis_ave5days	-8.71×10^2	7.42×10^5	−0.001	0.999
crisis_lag1	2.61	1.56	1.67	0.0942	crisis_lag1	2.19×10^2	1.23×10^5	0.002	0.999
hen_return_lag1	39.25	18.85	2.08	0.0373 *	hen_return_lag1	2.90×10^3	1.31×10^6	0.002	0.998

Notes: *** < 0.001 p-value; * < 0.05 p-value.

Table A2. Estimated LogR model result: All variables.

	All Variables								
	Moving Window Method					Expanding Window Method			
	Estimate	Std. Error	z-Value	Pr(>|z|)		Estimate	Std. Error	z-Value	Pr(>|z|)
(Intercept)	−4.56	0.21	−21.87	2×10^{-16} ***	(Intercept)	−5.48	1.73	−3.17	0.00153 **
crisis_ave5days	8.75	2.34	3.73	0.000190 ***	crisis_ave5days	7.30	2.58	2.83	0.00465 **
hen_return_lag5	7.31	2.21	3.31	0.000944 ***	hen_return_lag5	6.60	2.39	2.76	0.00582 **
hen_return_lag2	9.19	4.07	2.26	0.023885 *	gold_return_lag5	2.61	9.88	2.64	0.00835 **

Notes: *** < 0.001 p-value; ** < 0.01 p-value; * < 0.05 p-value.

References

1. Birol, F. The Impact of the Financial and Economic Crisis on Global Energy Investments. In Proceedings of the Energy, Policies and Technologies for Sustainable Economies, IAEE (International Association for Energy Economics) European Conference, Vienna, Italy, 7–10 September 2009.
2. Klopotan, I.; Zoroja, J.; Mesko, M. Early warning system in business, finance, and economics: Bibliometric and topic analysis. *Int. J. Eng. Bus. Manag.* **2018**, *10*, 1–12. [CrossRef]
3. Edison, J.H. Do indicators of financial crises work? An evaluation of an early warning system. *Int. J. Financ. Econ.* **2003**, *8*, 11–53. [CrossRef]
4. Kaminsky, L.G.; Lizondo, S.; Reinhart, M.C. Leading indicators of currency crises. *IMF Econ. Rev.* **1998**, *45*, 1–48. [CrossRef]
5. Kaminsky, L.G.; Reinhart, M.C. The twin crises: The causes of banking and balance-of-payments problems. *Am. Econ. Rev.* **1999**, *89*, 473–500. [CrossRef]
6. Lin, W.Y.; Hu, Y.H.; Tsai, C.F. Machine learning in financial crisis prediction: A survey. *IEEE Trans. Syst. Man. Cybern. Part C* **2012**, *42*, 421–436.
7. Chen, S.W.C.; Gerlach, R.; Lin, M.H.E.; Lee, W.C.W. Bayesian forecasting for financial risk management, pre and post the global financial crisis. *J. Forecast.* **2012**, *31*, 661–687. [CrossRef]
8. Bagheri, A.; Peyhani, M.H.; Akbari, M. Financial forecasting using ANFIS networks with quantum-behaved particle swarm optimization. *Expert Syst. Appl.* **2014**, *41*, 6235–6250. [CrossRef]
9. Niemira, P.M.; Saaty, L.T. An analytic network process model for financial-crisis forecasting. *Int. J. Forecast.* **2004**, *20*, 573–587. [CrossRef]
10. Chiang, W.C.; Enke, D.; Wu, T.; Wang, R.Z. An adaptive stock index trading decision support system. *Expert Syst. Appl.* **2016**, *59*, 195–207. [CrossRef]
11. Gorucu, B.F. Artificial neural network modeling for forecasting gas consumption. *Energy Source* **2004**, *26*, 299–307. [CrossRef]
12. Xie, W.; Yu, L.; Xu, S.Y.; Wang, S.Y. A New Method for Crude Oil Price Forecasting Based on Support Vector Machines. In *International Conference on Computational Science*; Springer: Berlin/Heidelberg, Germany, 2006; pp. 444–451.
13. Frankel, A.J.; Rose, K.A. Currency crashes in emerging markets: An empirical treatment. *J. Int. Econ.* **1996**, *41*, 351–366. [CrossRef]
14. Bae, K.H.; Karolyi, A.; Stulz, M.R. A new approach to measuring financial contagion. *Rev. Financ. Stud.* **2003**, *16*, 717–763. [CrossRef]
15. Sachs, J.; Tornell, A.; Velasco, A. Financial crises in emerging markets: The lessons from 1995. *Brook. Pap. Econ. Act.* **1996**, *27*, 147–199. [CrossRef]

16. Knedlik, T.; Schweinitz, V.G. Macroeconomic imbalances as indicators for debt crises in Europe. *J. Common Market Stud.* **2012**, *50*, 726–745. [CrossRef]
17. Bussiere, M.; Fratzscher, M. Towards a new early warning system of financial crises. *J. Int. Money Financ.* **2006**, *25*, 953–973. [CrossRef]
18. Xu, L.; Kinkyo, T.; Hamori, S. Predicting currency crises: A novel approach combining random forests and wavelet transform. *J. Risk Financ. Manag.* **2018**, *11*, 86. [CrossRef]
19. Saleh, N.; Casu, B.; Clare, A. Towards a new model for early warning signals for systemic financial fragility and near crises: An application to OECD. *Econom. Econom. Model Constr. Estim. Sel. J.* 2012. [CrossRef]
20. Ahn, J.J.; Oh, K.J.; Kim, T.Y.; Kim, D.H. Usefulness of support vector machine to develop an early warning system for financial crisis. *Expert Syst. Appl.* **2011**, *38*, 2966–2973. [CrossRef]
21. Sevim, C.; Oztekin, A.; Bali, O.; Gumus, S.; Guresen, E. Developing an early warning system to predict currency crises. *Eur. J. Oper. Res.* **2014**, *237*, 1095–1104. [CrossRef]
22. Lin, C.S.; Khan, A.H.; Chang, R.Y.; Wang, Y.C. A new approach to modeling early warning systems for currency crises: Can a machine-learning fuzzy expert system predict the currency crises effectively? *J. Int. Money Financ.* **2008**, *27*, 1098–1121. [CrossRef]
23. Chatzis, P.S.; Siakoulis, V.; Petropoulos, A.; Stavroulakis, E.; Vlachogiannakis, N. Forecasting stock market crisis events using deep and statistical machine learning techniques. *Expert Syst. Appl.* **2018**, *112*, 353–371. [CrossRef]
24. Bolbol, A.; Cheng, T.; Tsapakis, I.; Haworth, J. Inferring hybrid transportation modes from sparse GPS data using a moving window SVM classification. *Comput. Environ. Urban Syst.* **2012**, *36*, 526–537. [CrossRef]
25. Chou, J.S.; Ngo, N.T. Time series analytics using sliding window metaheuristic optimization-based machine learning system for identifying building energy consumption patterns. *Appl. Energy* **2016**, *177*, 751–770. [CrossRef]
26. Oh, K.J.; Kim, T.Y.; Kim, C. An early warning system for detection of financial crisis using financial market volatility. *Expert Syst.* **2006**, *23*, 63–125. [CrossRef]
27. Patel, A.S.; Sarkar, A. Crises in developed and emerging stock markets. *Financ. Anal. J.* **1998**, *54*, 50–61. [CrossRef]
28. Coudert, V.; Gex, M. Does risk aversion drive financial crises? Testing the predictive power of empirical indicators. *J. Empir. Financ.* **2008**, *15*, 167–184. [CrossRef]
29. Li, W.X.; Chen, C.S.C.; French, J.J. Toward an early warning system of financial crises: What can index futures and options tell us? *Q. Rev. Econ. Financ.* **2015**, *55*, 87–99. [CrossRef]
30. Chen, T.Q.; He, T. xgboost: eXtreme Gradient Boosting. 2019. Available online: http://cran.fhcrc.org/web/packages/xgboost/vignettes/xgboost.pdf (accessed on 1 November 2019).
31. Vapnik, V.N. An overview of statistical learning theory. *IEEE Trans. Neural Netw.* **1999**, *10*, 988–999. [CrossRef]
32. Ohlson, J. Financial ratios and the probabilistic prediction of bankruptcy. *J. Account. Res.* **1980**, *18*, 109–131. [CrossRef]
33. Breiman, L. Random forests. *Mach. Learn.* **2001**, *45*, 5–32. [CrossRef]
34. Tanaka, K.; Kinkyo, T.; Hamori, S. Random Forests-based Early Warning System for Bank Failures. *Econ. Lett.* **2016**, *148*, 118–121. [CrossRef]
35. Tanaka, K.; Higashide, T.; Kinkyo, T.; Hamori, S. Analyzing industry-level vulnerability by predicting financial bankruptcy. *Econ. Inq.* **2019**, *57*, 2017–2034. [CrossRef]
36. Werbos, P.J. Advanced forecasting methods for global crisis warning and models of intelligence. *Gen. Syst. Yearb.* **1977**, *22*, 25–38.
37. Friedman, J.H. Greedy function approximation: A gradient boosting machine. *Ann. Stat.* **2001**, *29*, 1189–1232. [CrossRef]
38. Erdogen, E.B. Prediction of bankruptcy using support vector machines: An application to bank bankruptcy. *J. Stat. Comput. Sim.* **2012**, *83*, 1543–1555. [CrossRef]
39. Yap, F.C.B.; Munuswamy, S.; Mohamed, B.Z. Evaluating company failure in Malaysia using financial ratios and logistic regression. *Asian J. Financ. Account.* **2012**, *4*, 330–344. [CrossRef]
40. Hothorn, T.; Everitt, B.S. *A Handbook of Statistical Analyses Using R*, 3rd ed.; CRC Press: Boca Raton, FL, USA, 2014.
41. Ho, T.K. Random decision forests. In Proceedings of the Third International Conference on Document Analysis and Recognition, Montreal, QC, Canada, 14–16 August 1995; Volume 1, pp. 278–282.

42. Bekkar, M.; Djemaa, K.H.; Alitouche, A.T. Evaluation measures for models assessment over imbalanced data sets. *J. Inf. Eng. Appl.* **2013**, *3*, 27–38.
43. Powers, D.M. Evaluation: From precision, recall and fmeasure to roc, informedness, markedness and correlation. *J. Mach. Learn. Technol.* **2011**, *2*, 37–63.
44. Diebold, F.X.; Mariano, R. Comparing predictive accuracy. *J. Bus. Econ. Stat.* **1995**, *13*, 253–265.

© 2020 by the authors. Licensee MDPI, Basel, Switzerland. This article is an open access article distributed under the terms and conditions of the Creative Commons Attribution (CC BY) license (http://creativecommons.org/licenses/by/4.0/).

Article

Influence of Fluctuations in Fossil Fuel Commodities on Electricity Markets: Evidence from Spot and Futures Markets in Europe

Tiantian Liu [1], Xie He [1], Tadahiro Nakajima [1,2] and Shigeyuki Hamori [1,*]

[1] Graduate School of Economics, Kobe University, Kobe 657-8501, Japan; tiantian.liu.econ@gmail.com (T.L.); kakyo1515@gmail.com (X.H.); nakajima.tadahiro@a4.kepco.co.jp (T.N.)
[2] The Kansai Electric Power Company, Incorporated, Osaka 530-8270, Japan
[*] Correspondence: hamori@econ.kobe-u.ac.jp; Tel.: +81-78-657-8501

Received: 4 March 2020; Accepted: 11 April 2020; Published: 13 April 2020

Abstract: Using a fresh empirical approach to time-frequency domain frameworks, this study analyzes the return and volatility spillovers from fossil fuel markets (coal, natural gas, and crude oil) to electricity spot and futures markets in Europe. In the time domain, by an approach developed by Diebold and Yilmaz (2012) which can analyze the directional spillover effect across different markets, we find natural gas has the highest return spillover effect on electricity markets followed by coal and oil. We also find that return spillovers increase with the length of the delivery period of electricity futures. In the frequency domain, using the methodology proposed by Barunik and Krehlik (2018) that can decompose the spillover effect into different frequency bands, we find most of the return spillovers from fossil fuels to electricity are produced in the short term while most of the volatility spillovers are generated in the long term. Additionally, dynamic return spillovers have patterns corresponding to the use of natural gas for electricity generation, while volatility spillovers are sensitive to extreme financial events.

Keywords: electricity; crude oil; natural gas; coal; spillover effects

1. Introduction

Electricity is traded in both spot and futures markets. Extensive research has been undertaken over the past two decades concerning the spot market for electricity (e.g., [1–4]). Electricity is a non-storable good and faces generation constraints, transmission constraints, and seasonal issues, which can cause a timing imbalance in trade and fluctuations in electricity prices in the spot market.

Electricity futures have often played two roles for investors. First, electricity futures are an essential indicator to predict future spot prices. Second, electricity futures provide investors who are willing to take positions in power markets an excellent risk management tool for reducing their risk exposure, by reducing the operational risks caused by high volatility in the electricity spot market. For this reason, electricity futures trading is more extensive than spot trading. There are also many studies concerning the electricity futures market (e.g., [5–8]).

Approximately half of the world's electricity is generated by fossil fuels such as crude oil, natural gas, and coal. According to the BP Statistical Review of World Energy 2019 [9], in 2018, approximately 40.5% of electricity was produced by oil, gas, and coal sources in Europe, which also necessarily implies the possibility of the spillover effects across the fossil fuel markets and electricity market. Additionally, in Huisman and Kilic's [5] study, they state that although electricity cannot be stored directly, it can be stored in the sense that fossil fuels are storable. The electricity producer who wants to sell an electricity futures contract can either purchase the amount of fossil fuels in the spot market and store it until the delivery period to fulfill the delivery agreement or purchase the fossil fuel futures instead of storing them directly. For this reason, the price of electricity futures may be related to the storage cost of fossil fuels. On the other hand, in electricity spot market, according to the Mosquera-López and Nursimulu's [10] research, unlike futures markets, the electricity spot prices are more determined by renewable energy infeed and electricity demand but not the price of fossil fuels such as natural gas, coal, and carbon. Thus, our research aims to examine if there are some difference in the spillover effects of fossil fuel commodity markets on the electricity market and futures in Europe. Additionally, given that there are many types of electricity futures contracts in terms of their delivery periods, we further examine whether there are different spillover effects between fossil fuel commodities and electricity futures with monthly, quarterly, and yearly delivery periods. To the best of our knowledge, although there are several papers discussing the relationship between the fossil fuels and electricity futures market, the relationship between fossil fuels and the price of electricity futures with different delivery periods have not been studied in previous research yet. However, it will help electricity market investors to fully understand the information transmission between electricity markets and fossil fuel markets to make their portfolio optimization and hedge strategy better and more comprehensive.

Many studies have investigated the interdependent relationship between fossil fuel commodities and electricity markets. For example, Emery and Liu [11] studied the relationship between the prices of electricity futures and natural gas futures and found a cointegration between California–Oregon Border and Palo Verde electricity futures and natural gas futures. Mjelde and Bessler [12] used a vector error correction model to analyze the relationship between electricity spot prices and electricity-generating fuel sources (natural gas, crude oil, coal, and uranium) in the US. The authors found that the peak electricity price influences the natural gas price in contemporaneous time, while in the long term, apart from uranium, fuel source prices affect the electricity price. Based on the VECM model, Furió and Chuliá [13] analyzed the volatility and price linkages between the Spanish electricity market, Brent crude oil, and Zeebrugge (Belgium) natural gas. Natural gas and crude oil were seen to have an essential influence in the Spanish electricity market, with particular causality from the fossil fuel (Brent crude oil and Zeebrugge natural gas) markets to the Spanish electricity forward market.

Generalized autoregressive conditional heteroskedasticity (GARCH)-type models have also been used to study the relationship between electricity and fossil fuel markets. Serletis and Shahmoradi [14] used the GARCH-M model to investigate the linkages between natural gas and electricity prices and volatilities in Alberta, Canada, and found a bidirectional causality. Using the BEKK-GARCH model, Green et al. [15] measured the strength of volatility spillovers in the electricity market in Germany caused by volatility in natural gas, coal, and carbon emission markets. According to the results, during the sample period, natural gas and coal produced non-negligible spillovers while carbon emission markets caused spillover effects from 2011 to 2014.

The return spillover is defined as the cross-market correlation between price changes (returns) allowing for one market's price fluctuation to affect the other market's direction with a lag. On the other hand, the volatility spillover could capture the correlation between the size of price changes. That is, if one market becomes riskier over a period of time (which implies bigger price fluctuations) the riskiness of the interrelated market will also change. Measuring the return spillover could allow investors to evaluate the investment trends of markets while measuring the volatility spillover could enable investors to evaluate the risk information across markets. The information across markets provide useful insights into the portfolio diversification of investment, hedging strategies, and risk management for financial agents.

In this study, we analyze the return and volatility spillovers from the fossil fuel market commodities of natural gas, coal, and crude oil, to electricity spot and futures markets in Europe, by using two new empirical methods in the time-frequency domain: (1) the Diebold–Yilmaz approach; and (2) the Barunik and Krehlik methodology. The Diebold–Yilmaz approach was proposed in Diebold and Yilmaz [16] and developed in Diebold and Yilmaz [17]. In the Diebold–Yilmaz approach, the spillover index can be constructed based on the variance decomposition of forecast error, which allows for the study of the spillover effect with a fixed investment horizon in a quantitative way. However, investors have to consider different investment horizons when they make investment decisions because shocks may have different effects in the short and long term. For this reason, Barunik and Krehlik [18] used the Fourier transform to convert the Diebold–Yilmaz approach into the frequency domain so that the spillover index could be decomposed at different frequencies. Moreover, we can also obtain the time-varying spillover effect by using the moving window method. Many recent studies have used this method to study spillover effects and connectedness between assets, such as Singh et al. [19], Kang and Lee [20], and Malik and Umar [21]. In a more recent study, Lovcha and Perez-Laborda [22] used the Diebold–Yilmaz approach and the Barunik and Krehlik methodology to analyze the volatility connectedness between Henry Hub natural gas and West Texas Intermediate (WTI) crude oil in the time and frequency domains. Lau et al. [23] used the E-GARCH model and the Barunik and Krehlik methodology approach to investigate the return and volatility spillover effects among white precious metals, gold, oil, and global equity.

The remainder of this paper is structured as follows: Section 2 introduces the empirical method used in this research, while Section 3 introduces the data used and preliminary analysis. Section 4 explains and analyzes the empirical results, and Section 5 presents the conclusions.

2. Method Framework

Diebold and Yilmaz [17] proposed an approach for measuring spillover in the generalized vector autoregression framework using the concept of connectedness, which built on the generalized forecast error variance decomposition (GFEVD) of a Vector Autoregressive (VAR) model with p lags.

$$y_t = \sum_{i=1}^{p} \Phi_i y_{t-i} + \varepsilon_t, \tag{1}$$

where y_t is an $N \times 1$ vector of observed variables at time t, Φ is the $N \times N$ coefficient matrices, and error vector ε_t i.i.d $\sim (0, \Sigma)$ with covariance matrix Σ is possibly non-diagonal.

The VAR process can also be represented as the following Moving Average (∞) representation Equation (2), assuming the roots of $|\Phi(z)|$ lie outside the unit-circle:

$$y_t = \Psi(L) \varepsilon_t, \tag{2}$$

where $\Psi(L)$ is an $N \times N$ coefficient matrix of infinite lag polynomials. Since the order of variables in the VAR system may have an influence on the impulse response or variance decomposition results, Diebold and Yilmaz [17] modified the generalized VAR framework of Koop et al. [24] and Pesaran and Shin [25] to ensure the variance decomposition's independence of ordering. Under such a framework, the H-step-ahead generalized forecast error variance decomposition (GFEVD) can be presented in the form of Equation (3):

$$\theta^H_{jk} = \frac{\sigma^{-1}_{kk} \sum_{h=0}^{H} \left((\Psi_h \Sigma)_{jk} \right)^2}{\sum_{h=0}^{H} \left(\Psi_h \Sigma \Psi'_h \right)_{jj}}, \tag{3}$$

where Ψ_h stands for an $N \times N$ coefficient matrix of polynomials at lag h, and $\sigma_{kk} = (\Sigma)_{kk}$. The θ^H_{jk} represents H-steps ahead forecast error variance of the element j which is contributed by the k-th variable of the VAR system. To make the sum of the elements in each row of the generalized forecast error variance decomposition (GFEVD) equal to 1, each entry is standardized by the row sum as:

$$\tilde{\theta}^H_{jk} = \frac{\theta^H_{jk}}{\sum_{k=1}^{N} \theta^H_{jk}}, \tag{4}$$

The $\widetilde{\theta}_{jk}^{H}$ is also defined as the Pairwise Spillover from k to j over the horizon H, which can measure the spillover effect from market k to market j. Meanwhile, there are also directional spillovers as defined by Diebold and Yilmaz:

Directional Spillovers (From): $S_{k\leftarrow \cdot}^{H} = 100 \times \dfrac{\sum_{\substack{j=1 \\ j \neq k}}^{N} \widetilde{\theta}_{kj}^{H}}{N}$, the directional spillovers (From) measure the spillovers from all other markets to market k.

The above-defined measures of spillovers are summarized in Table 1.

Table 1. The spillover table of the Diebold–Yilmaz approach (2012).

			Spillover Results			
	y_1	y_2	...	y_N	From others	
y_1	$S_{1\leftarrow 1}$	$S_{1\leftarrow 2}$	...	$S_{1\leftarrow N}$	$S_{1\leftarrow \cdot}^{H} = \frac{1}{N}\sum_{j=1}^{N} S_{1\leftarrow j}, j \neq 1$	
y_2	$S_{2\leftarrow 1}$	$S_{2\leftarrow 2}$	...	$S_{2\leftarrow N}$	$S_{2\leftarrow \cdot}^{H} = \frac{1}{N}\sum_{j=1}^{N} S_{2\leftarrow j}, j \neq 2$	
⋮	⋮	⋮	⋱	⋮	⋮	
y_N	$S_{N\leftarrow 1}$	$S_{N\leftarrow 2}$	...	$S_{N\leftarrow N}$	$S_{N\leftarrow \cdot}^{H} = \frac{1}{N}\sum_{j=1}^{N} S_{N\leftarrow j}, j \neq N$	

Note: adapted from "On the Network Topology of Variance Decompositions: Measuring the Connectedness of Financial Firms" [26].

Barunik and Krehlik [1] proposed a methodology that could decompose the spillover index in the Diebold–Yilmaz approach on the specific frequency bands.

First and foremost, by the application of the Fourier transform, spillovers in the frequency domain can be measured. Through a Fourier transform of the coefficients to obtain a frequency response function Ψ_h: $\Psi(e^{-i\omega}) = \sum_h e^{-i\omega h} \Psi_h$, where $i = \sqrt{-1}$. The generalized causation spectrum over frequencies $\omega \in (-\pi, \pi)$ can be defined as:

$$(f(\omega))_{jk} \equiv \dfrac{\sigma_{kk}^{-1} \left| \left(\Psi(e^{-i\omega}) \Sigma \right)_{jk} \right|^2}{\left(\Psi(e^{-i\omega}) \Sigma \Psi'(e^{+i\omega}) \right)_{jj}} \tag{5}$$

It is crucial to note that $(f(\omega))_{jk}$ represents the contribution of the k-th variable to the portion of the spectrum of the j-th variable at a given frequency ω. To find the generalized decomposition of variance to different frequencies, $(f(\omega))_{jk}$ can be weighted by the frequency share of the variance of the j-th variable. The weighting function can be defined as:

$$\Gamma_j(\omega) = \dfrac{\left(\Psi(e^{-i\omega}) \Sigma \Psi'(e^{+i\omega}) \right)_{jj}}{\frac{1}{2\pi} \int_{-\pi}^{\pi} \left(\Psi(e^{-i\lambda}) \Sigma \Psi'(e^{+i\lambda}) \right)_{jj} d\lambda} \tag{6}$$

Equation (6) gives a presentation of the power of the j-th variable at a given frequency, the sum of the frequencies to a constant value of 2π. Meanwhile, it is important to note that the generalized factor spectrum is the squared coefficient of weighted complex numbers and is a real number when the Fourier transform of the impulse is a complex number value. Therefore, we can set up the frequency band $d = (a,b)$: $a, b \in (-\pi, \pi)$, $a < b$.

The GFEVD under the frequency band d is:

$$\theta_{jk}(d) = \dfrac{1}{2\pi} \int_a^b \Gamma_j(\omega) (f(\omega))_{jk} d\omega \tag{7}$$

This also needs to be normalized; the scaled GFEVD on the frequency band $d = (a,b)$: $a, b \in (-\pi, \pi)$, $a < b$ can be defined as:

$$\widetilde{\theta}_{jk}(d) = \frac{\theta_{jk}(d)}{\sum_k \theta_{jk}(\infty)} \quad (8)$$

We can define $\widetilde{\theta}_{jk}(d)$ as the *pairwise spillover* on a given frequency band d. Meanwhile, given the *total spillover* proposed by Diebold and Yilmaz [16], it is possible to define the *total spillover* on the frequency band d. Similarly, there are also directional spillovers in the frequency domain:

Frequency Directional Spillovers (From): $S^{\mathcal{F}}_{k \leftarrow \cdot}(d) = 100 \times \dfrac{\sum_{\substack{j=1 \\ j \neq k}}^{N} \widetilde{\theta}_{kj}(d)}{N}$, the frequency directional spillovers (From) represents the spillovers from all other markets to market k on the frequency band d.

Similar to Table 1, the above-defined measures of spillovers on the frequency band d are summarized in Table 2.

Table 2. The spillover table of the Barunik–Krehlik methodology (2018).

			Spillover Results		
	y_1	y_2	...	y_N	From others
y_1	$S^{\mathcal{F}}_{1\leftarrow 1}(d)$	$S^{\mathcal{F}}_{1\leftarrow 2}(d)$	...	$S^{\mathcal{F}}_{1\leftarrow N}(d)$	$S^{\mathcal{F}}_{1\leftarrow \cdot}(d) = \frac{1}{N}\sum_{j=1}^{N} S^{\mathcal{F}}_{1\leftarrow j}(d), j \neq 1$
y_2	$S^{\mathcal{F}}_{2\leftarrow 1}(d)$	$S^{\mathcal{F}}_{2\leftarrow 2}(d)$	...	$S^{\mathcal{F}}_{2\leftarrow N}(d)$	$S^{\mathcal{F}}_{2\leftarrow \cdot}(d) = \frac{1}{N}\sum_{j=1}^{N} S^{\mathcal{F}}_{2\leftarrow j}(d), j \neq 2$
⋮	⋮	⋮	⋱	⋮	⋮
y_N	$S^{\mathcal{F}}_{N\leftarrow 1}(d)$	$S^{\mathcal{F}}_{N\leftarrow 2}(d)$	...	$S^{\mathcal{F}}_{N\leftarrow N}(d)$	$S^{\mathcal{F}}_{N\leftarrow \cdot}(d) = \frac{1}{N}\sum_{j=1}^{N} S^{\mathcal{F}}_{N\leftarrow j}(d), j \neq N$

Note: adapted from "On the Network Topology of Variance Decompositions: Measuring the Connectedness of Financial Firms" [26].

3. Data and Preliminary Analysis

In this study, we use data from three fossil fuels markets: natural gas, coal, and crude oil; and four electricity markets: the electricity spot market and the monthly, quarterly, and yearly electricity futures markets. We employ daily data for the period from 2 January 2007, to 2 January 2019, with a total of 3019 observations. These data have been converted local currencies into euros at the daily exchange rate. All variables we used are listed in Table 3.

Table 3. Variables in the model.

Variable	Data	Data Source
Ele_Spot	EPEX Germany Baseload Spot	Bloomberg
Ele_Fut_Mon	EEX Germany Baseload Monthly Futures	Bloomberg
Ele_Fut_Qr	EEX Germany Baseload Quarterly Futures	Bloomberg
Ele_Fut_Yr	EEX Germany Baseload Yearly Futures	Bloomberg
Gas	ICE UK Natural Gas Monthly Futures	Bloomberg
Coal	ICE Rotterdam Coal Monthly Futures	Bloomberg
Oil	ICE Brent Crude Oil Monthly Futures	Bloomberg

For the fossil fuels markets, we used a representative price of natural gas, coal, and crude oil futures markets in Europe. First, we used the United Kingdom (UK) National Balancing Point (NBP) natural gas futures to represent the natural gas market. The NBP natural gas market is the oldest natural gas market in Europe and is widely used as a leading benchmark for the wholesale gas market in Europe. Second, we used the Rotterdam coal futures to represent the coal market, which is financially settled based upon the price of coal delivered into the Amsterdam, Rotterdam, and Antwerp regions of the Netherlands. The futures contract is cash-settled against the API 2 Index, which is the benchmark price reference for coal imported to Northwest Europe. Third, we used the Brent crude oil futures to represent the crude oil market. The Brent crude oil market is one of the world's most liquid crude oil markets. The price of Brent crude oil is the benchmark for African, European, and Middle Eastern crude. All futures mentioned above are traded on the Intercontinental Exchange (ICE) Futures Europe commodities market. The price of natural gas quoted in EUR per therm, the price of coal quoted in EUR per metric ton, and the price of crude oil quoted in EUR per barrel.

In Europe, electricity futures markets are offered in the European Energy Exchange (EEX). The EEX is a central European electric power exchange established in August 2000 and has become the leading energy exchange in Europe with more than 200 trading participants from 19 countries. In 2008, the power spot markets EEX and Powenext merged to create the European Power Exchange (EPEX SPOT) to offer spot markets in electricity. In this study, we used the spot, monthly futures, quarterly futures, and yearly futures markets of the Physical Electricity Index (Phelix) Baseload, which is the reference in Germany and majority of Europe. The electricity spot commodity is traded in the EPEX SPOT, and futures commodities are traded in the EEX. The prices of electricity spot and futures quoted in EUR per megawatt-hour (Mwh).

The closing prices of all variables are plotted in Figure 1. As shown, we find that the variation in electricity spot prices is more extreme, with more spikes and jumps than other commodities. It is clear that electricity futures and fossil fuel prices show a similar pattern.

In this study, we calculated the first order logged differences in prices as the returns and extracted the conditional variance series by fitting the AR-GARCH model as the volatilities. The descriptive statistics for the returns and volatilities of the variables are given in Table 4, where we can see, as expected, that the electricity spot market is the most volatile. The skewness value of the return of coal futures is negative and others are positive. Meanwhile, quarterly electricity futures have the highest skewness, indicating that they have the most extreme gains.

Table 4. Descriptive statistics of the return and volatility series.

	Spot	Fut_Mon	Fut_Qr	Fut_Yr	Gas	Coal	Oil
			Descriptive Statistics of the Return				
Mean	0.0002	0.0000	0.0001	−0.0000	0.0001	0.0001	0.0000
Minimum	−2.0000	−0.2429	−0.2002	−0.09366	−0.1334	0.2179	−0.1094
Maximum	2.3785	0.4042	0.3422	0.1535	0.3589	0.1751	0.1718
Std. Dev.	0.2537	0.0315	0.0204	0.0118	0.0296	0.0161	0.0211
Skewness	0.6253	2.415	3.1200	0.5691	1.8429	−0.4293	0.0590
Kurtosis	11.4001	38.3887	72.8856	16.7734	22.1089	35.6286	7.3911
JB	9072.9 ***	160,472 ***	619,264 ***	24,026 ***	47,642 ***	134,014 ***	2427.2 ***
ADF	−56.9 ***	−37.7 ***	−36.9 ***	−38.0 ***	−39.8 ***	−39.1 ***	−39.9 ***
PP	−123.9 ***	−52.1 ***	−51.6 ***	−50.9 ***	−52.2 ***	−53.2 ***	−59.1 ***
			Descriptive Statistics of the Volatility				
Mean	0.0589	0.0010	0.0003	0.0001	0.0001	0.0003	0.0004
Minimum	0.0160	0.0006	0.0000	0.0000	0.0009	0.0001	0.0001
Maximum	0.9180	0.0020	0.0068	0.0023	0.0136	0.0018	0.0048
Std. Dev.	0.0589	0.0003	0.0005	0.0002	0.0012	0.0002	0.0005
Skewnesss	5.3783	1.3262	6.3463	4.0463	4.0977	2.9271	3.5460
Kurtosis	52.2126	4.0317	58.6978	30.5627	29.5249	12.4334	20.0091
JB	319,207 ***	1018.8 ***	410,502 ***	103,802 ***	96,952 ***	15,505 ***	42,720 ***
ADF	−9.7 ***	−3.4 **	−10.3 ***	−5.6 ***	−7.0 ***	−3.0 **	−3.5 ***
PP	−11.7 ***	−3.3 **	−12.6 ***	−7.8 ***	−10.2 ***	−3.3 **	−3.9 ***

Note: Spot, Fut_Mon, Fut_Qr, Fut_Yr, Gas, Coal, Oil refer to EPEX Germany Baseload Spot, EEX Germany Baseload Monthly Futures, EEX Germany Baseload Quarterly Futures, EEX Germany Baseload Yearly Futures, ICE UK Natural Gas Monthly Futures, ICE Rotterdam Coal Monthly Futures, ICE Brent Crude Oil Monthly Futures, respectively. JB: Jarque and Bera Test (1980); ADF: Augmented Dickey and Fuller Unit Root Test (1979); PP: Phillips and Perron Unit Root Test (1988); *, ** and *** denote rejection of the null hypothesis at 10%, 5% and 1% significance levels, respectively.

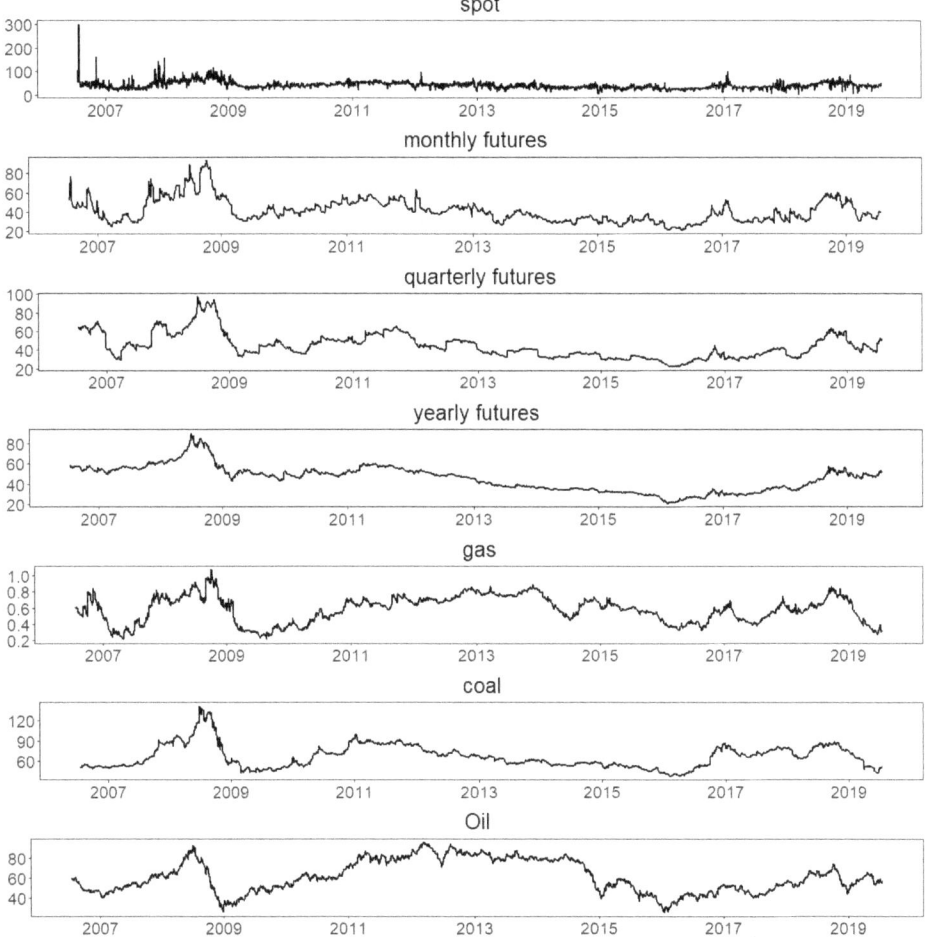

Figure 1. Time-variations of the price series. Note: spot: EPEX Germany Baseload Spot; monthly futures: EEX Germany Baseload Monthly Futures; quarterly futures: EEX Germany Baseload Quarterly Futures; yearly futures: EEX Germany Baseload Yearly Futures; gas: ICE UK Natural Gas Monthly Futures; coal: ICE Rotterdam Coal Monthly Futures; oil: ICE Brent Crude Oil Monthly Futures; the units of electricity spot, monthly futures, quarterly futures, and yearly futures are EUR/Mwh; the units of gas, coal, oil are EUR/therm, EUR/metric ton, and EUR/barrel, respectively.

All of the returns and volatilities have a kurtosis value that is significantly higher than three, implying that the distribution of returns and volatilities will show tails that are peaked and fat. The highest kurtosis value is for quarterly electricity futures both in returns and volatilities.

The results of the Jarque–Bera (JB) test show that all returns and volatilities reject normality at the 1% significance level. Finally, since the Diebold–Yilmaz approach is based on the VAR model, the data should be stationary. From the results of the Augmented Dickey–Fuller (ADF) unit root test and the Philips and Pearron (PP) unit root test, the null hypothesis that of each variable being nonstationary is rejected at the 1% or 5% significance level for all cases.

4. Empirical Results and Discussion

First, we used a four-variable VAR model with four different sets of variables (four sets of returns and four sets of volatilities): The first set is $y_t = (\text{Ele\_Spot}_t, \text{Gas}_t, \text{Coal}_t, \text{Oil}_t)$, the second set is $y_t = (\text{Ele\_Fut\_Mon}_t, \text{Gas}_t, \text{Coal}_t, \text{Oil}_t)$, the third set is $y_t = (\text{Ele\_Fut\_Qr}_t, \text{Gas}_t, \text{Coal}_t, \text{Oil}_t)$, and the fourth set is $y_t = (\text{Ele\_Fut\_Yr}_t, \text{Gas}_t, \text{Coal}_t, \text{Oil}_t)$. For the four sets of returns, according to the Schwarz Criterion (SC), the lag length of the VAR model is 4 in the first set and 1 in the other sets, while in the four sets of volatilities, the lag length of the VAR model is 2 in the first set and 1 in the other sets. Subsequently, the Diebold–Yilmaz approach, which is based on the generalized variance decomposition, is applied to the VAR models to assess the direction and intensity of the spillover index across the selected variables in the time domain.

Second, with the assistance of the Barunik and Krehik [18] methodology and following the methods of Toyoshima and Hamori [27], the spillover indexes obtained by the Diebold–Yilmaz approach were decomposed into three frequency bands: the high frequency, 'Frequency H', roughly corresponding to 1–5 days; the medium frequency, 'Frequency M', roughly corresponding to 5–21 days; and the low frequency, 'Frequency L', roughly corresponding to 21 days to infinity.

Finally, as noted by Barunik and Krehik [18], the methodology does not work if the forecasting horizon (H) < 100; therefore, the 100-days ahead forecasting horizon (H) for generalized variance decomposition was used in this study.

With the assistance of the Diebold–Yilmaz approach, Table 5 shows the return spillover effects from the three fossil fuels to the electricity spot market and the three types of futures markets. In addition, the last column of the table labelled 'Directional Spillover (From)' shows the total spillover effect from all fossil fuel commodities combined on the electricity spot market.

Table 5. The return spillover results of the Diebold–Yilmaz approach (2012).

	Return Spillovers				
	Electricity Spot				
	Ele_Spot	Gas	Coal	Oil	From
Ele_Spot	99.578	0.205	0.118	0.099	0.106
	Electricity Monthly Futures				
	Ele_Fut_Mon	Gas	Coal	Oil	From
Ele_Fut_Mon	93.738	4.549	1.348	0.364	1.565
	Electricity Quarterly Futures				
	Ele_Fut_Qr	Gas	Coal	Oil	From
Ele_Fut_Qr	87.512	8.142	2.719	1.627	3.122
	Electricity Yearly Futures				
	Ele_Fut_Yr	Gas	Coal	Oil	From
Ele_Fut_Yr	77.674	10.015	7.176	5.136	5.581

Note: Ele_Spot, Ele_Fut_Mon, Ele_Fut_Qr, Ele_Fut_Yr, Gas, Coal, Oil refer to EPEX Germany Baseload Spot, EEX Germany Baseload Monthly Futures, EEX Germany Baseload Quarterly Futures, EEX Germany Baseload Yearly Futures, ICE UK Natural Gas Monthly Futures, ICE Rotterdam Coal Monthly Futures, ICE Brent Crude Oil Monthly Futures, respectively.

Table 5 highlights several important findings. First, as shown in each row of the table, for both spot and futures markets, natural gas is the largest contributor to the forecast error variance decomposition (FEVD) of electricity return, which further implies that natural gas has the most influence on the electricity return. Meanwhile, among the three fossil fuels, crude oil has the least influence upon the electricity return. This may be due to electricity production from one specific fuel and storage costs. According to the BP Statistical Review of World Energy 2019 [9], in Europe, from 2007 to 2018, almost 25.35% of electricity was produced from coal, followed by 18.88% from natural gas, and 2.02% from crude oil. This explains why crude oil has the least influence upon the electricity return. Fama and

French [28] state that a trader can offset the risk of positions they have in a forward contract by holding a long or short inventory in the underlying commodity. Huisman and Kilic [5] note that although electricity cannot (yet) be stored directly, it can be stored in the sense that fossil fuels are storable. When an electricity producer sells an electricity futures contract, they have two choices: (1) purchase the amount of fossil fuels in the spot market and store it until the delivery period to fulfill the delivery agreement or; (2) purchase the fossil fuel futures instead of storing them directly. If the fuels have a high storage cost, the electricity producers would prefer to hold a futures contract on the fuels rather than purchase them in the spot market and store them directly. Compared to coal and crude oil, natural gas has higher storage costs and, therefore, the electricity producer would prefer to purchase natural gas futures. This further explains the high connectedness of electricity and natural gas futures markets. Based on high demand and high storage costs, it is not difficult to understand why natural gas futures have the most influence on the returns of electricity.

Second, it is interesting to note that the most significant return spillover effects are seen in yearly electricity futures, followed by the spillover effect to quarterly futures, then monthly futures. The return spillovers from fossil fuel commodities is the least in the electricity spot market. Under the assumption of storage costs mentioned earlier, when an electricity producer sells an electricity futures contract with a long delivery period, they would prefer to purchase futures contracts on underlying fuel commodities, rather than purchase them in the spot market and store until the delivery period. This implies that the return spillovers from fuel futures to electricity futures with a long delivery period are higher than to electricity futures with a short delivery period. Mosquera-López and Nursimulu [10] explored the drivers of German electricity prices in spot and futures markets and found that spot prices are determined by renewable energy infeed and electricity demand, while in futures markets prices are determined by the price of fossil fuels such as natural gas, coal, and carbon. This may explain why the return spillovers from fossil fuel commodities to electricity futures are higher than to electricity spot prices.

The return spillover results of Barunik and Krehil [16], which decompose the Diebold–Yilmaz spillover indexes into three different frequencies, are presented in Table 6. The spillovers from fossil fuel commodities to electricity markets are highest in the high frequency, followed by the medium frequency and the low frequency. This indicates that return shocks are transmitted from fossil fuel markets to electricity markets within only one week.

Table 7 shows the volatility spillover effects from the threVe fossil fuel commodities to the electricity spot market and the three types of futures. The results are unusual; among the three fossil fuel markets, natural gas has the highest volatility spillovers to electricity spot markets (0.691%), monthly (9.539%), and quarterly futures (3.565%), but not to yearly futures (0.145%). Additionally, oil exhibits the weakest volatility spillovers to all electricity markets (0.098%, 0.417%, and 0.211%, respectively), again, with the exception of yearly futures (1.835%).

In addition, in contrast to the return spillover results, the volatility spillovers from natural gas to monthly electricity futures (9.539%) is higher than for electricity futures with a longer delivery period. However, crude oil and coal transmit the highest volatility spillovers to yearly futures, while the spillovers to monthly futures are higher than to quarterly futures. Finally, with the exception of natural gas, both volatility spillovers and return spillovers are higher for electricity futures than for the electricity spot market.

Table 8 displays the volatility spillover results in the frequency domain. In contrast to the results in Table 6, the total volatility spillover is higher in the low frequency than in the high frequency. This indicates that the transmission of volatility shocks from fossil fuel markets to electricity markets is slower than that of return spillovers. The transmitted shocks from fossil fuels have long-lasting effects on electricity market volatility.

Table 6. The return spillover results of the Barunik–Krehlik methodology (2018).

Return Spillovers				
Electricity Spot				
Frequency H 1–5 Days				
Ele_Spot	Gas	Coal	Oil	From
Ele_Spot 97.182	0.193	0.116	0.084	0.098
Frequency M 5–21 Days				
Ele_Spot 1.902	0.009	0.002	0.011	0.005
Frequency L > 21 Days				
Ele_Spot 0.494	0.003	0.001	0.004	0.002
Electricity Monthly Futures				
Frequency H 1–5 Days				
Ele_Fut_Mon	Gas	Coal	Oil	From
Ele_Fut_Mon 74.381	2.913	0.961	0.250	1.031
Frequency M 5–21 Days				
Ele_Fut_Mon 14.234	1.191	0.283	0.083	0.389
Frequency L > 21 Days				
Ele_Fut_Mon 5.123	0.445	0.104	0.031	0.145
Electricity Quarterly Futures				
Frequency H 1–5 Days				
Ele_Fut_Qr	Gas	Coal	Oil	From
Ele_Fut_Qr 68.856	5.671	1.974	1.149	2.199
Frequency M 5–21 Days				
Ele_Fut_Qr 13.703	1.803	0.545	0.349	0.674
Frequency L > 21 Days				
Ele_Fut_Qr 4.953	0.668	0.200	0.129	0.249
Electricity Yearly Futures				
Frequency H 1–5 Days				
Ele_Fut_Yr	Gas	Coal	Oil	From
Ele_Fut_Yr 60.862	7.287	5.117	3.450	3.964
Frequency M 5–21 Days				
Ele_Fut_Yr 12.348	1.993	1.505	1.231	1.182
Frequency L > 21 Days				
Ele_Fut_Yr 4.464	0.734	0.553	0.455	0.436

Note: Ele_Spot, Ele_Fut_Mon, Ele_Fut_Qr, Ele_Fut_Yr, Gas, Coal, Oil refer to EPEX Germany Baseload Spot, EEX Germany Baseload Monthly Futures, EEX Germany Baseload Quarterly Futures, EEX Germany Baseload Yearly Futures, ICE UK Natural Gas Monthly Futures, ICE Rotterdam Coal Monthly Futures, ICE Brent Crude Oil Monthly Futures, respectively.

Table 7. The volatility spillover results of the Diebold–Yilmaz approach (2012).

Volatility Spillovers				
Electricity Spot				
	Ele_Spot	Gas	Coal	Oil	From
Ele_Spot	98.804	0.691	0.407	0.098	0.299
Electricity Monthly Futures				
	Ele_Fut_Mon	Gas	Coal	Oil	From
Ele_Fut_Mon	87.992	9.539	2.052	0.417	3.002
Electricity Quarterly Futures				
	Ele_Fut_Qr	Gas	Coal	Oil	From
Ele_Fut_Qr	95.585	3.565	0.639	0.211	1.104
Electricity Yearly Futures				
	Ele_Fut_Yr	Gas	Coal	Oil	From
Ele_Fut_Yr	88.839	0.145	9.181	1.835	2.790

Note: Ele_Spot, Ele_Fut_Mon, Ele_Fut_Qr, Ele_Fut_Yr, Gas, Coal, Oil refer to EPEX Germany Baseload Spot, EEX Germany Baseload Monthly Futures, EEX Germany Baseload Quarterly Futures, EEX Germany Baseload Yearly Futures, ICE UK Natural Gas Monthly Futures, ICE Rotterdam Coal Monthly Futures, ICE Brent Crude Oil Monthly Futures, respectively.

Table 8. The volatility spillover results of the Barunik–Krehlik methodology (2018).

Volatility Spillovers					
Electricity Spot					
Frequency H 1–5 Days					
	Ele_Spot	Gas	Coal	Oil	From
Ele_Spot	9.671	0.003	0.003	0.002	0.002
Frequency M 5–21 Days					
Ele_Spot	35.053	0.042	0.009	0.010	0.015
Frequency L > 21 Days					
Ele_Spot	54.079	0.646	0.395	0.086	0.282
Electricity Monthly Futures					
Frequency H 1–5 Days					
	Ele_Fut_Mon	Gas	Coal	Oil	From
Ele_Fut_Mon	0.297	0.009	0.010	0.009	0.007
Frequency M 5–21 Days					
Ele_Fut_Mon	0.852	0.029	0.029	0.026	0.021
Frequency L > 21 Days					
Ele_Fut_Mon	86.843	9.502	2.012	0.382	2.974
Electricity Quarterly Futures					
Frequency H 1–5 Days					
	Ele_Fut_Qr	Gas	Coal	Oil	From
Ele_Fut_Qr	9.962	0.195	0.000	0.014	0.052
Frequency M 5–21 Days					
Ele_Fut_Qr	24.567	0.536	0.001	0.034	0.143
Frequency L > 21 Days					
Ele_Fut_Qr	61.057	2.834	0.638	0.164	0.909
Electricity Yearly Futures					
Frequency H 1–5 Days					
	Ele_Fut_Yr	Gas	Coal	Oil	From
Ele_Fut_Yr	4.405	0.002	0.018	0.001	0.005
Frequency M 5–21 Days					
Ele_Fut_Yr	11.985	0.005	0.050	0.003	0.015
Frequency L > 21 Days					
Ele_Fut_Yr	72.448	0.138	9.113	1.831	2.771

Note: Ele_Spot, Ele_Fut_Mon, Ele_Fut_Qr, Ele_Fut_Yr, Gas, Coal, Oil refer to EPEX Germany Baseload Spot, EEX Germany Baseload Monthly Futures, EEX Germany Baseload Quarterly Futures, EEX Germany Baseload Yearly Futures, ICE UK Natural Gas Monthly Futures, ICE Rotterdam Coal Monthly Futures, ICE Brent Crude Oil Monthly Futures, respectively.

It is well established that financial markets can be affected by extreme events such as financial crises, price fluctuations, and market turmoil. For this reason, this section of our paper aims to capture the dynamics of spillover, in particular to investigate how spillovers from the fossil fuel market to the electricity market may change under extreme events. To accomplish this, a moving windows analysis with a 400-day window length was applied.

Figure 2 presents the dynamic directional return spillover from all fossil fuel commodities to electricity markets, measured by the Diebold–Yilmaz approach (2012) and the Barunik–Krehlik methodology (2018). The pairwise directional return spillovers are shown in Figure A1 in the Appendix A. In Figure 2, the solid line refers to the time-varying spillover of the Diebold–Yilmaz approach (2012), the long-dash line refers to spillover in the high frequency (1–5 days), the dotted line refers to spillover in the medium frequency (5–21 days), and the two-dash line refers to spillover in the low frequency (over 22 days). The results in Figure 2 indicate various characteristics.

First, the spillover effect of all three fossil fuel commodities to yearly electricity futures vary from 0% to 8.7%, followed by quarterly futures (0% to 6.3%), monthly futures (0% to 5.8%), and the spot market (0% to 2.8%). According to the results, it can be concluded that the return spillovers from fossil fuel commodities to electricity futures with a long delivery period are higher than to those with a short delivery period and that the return spillover effect is lowest in the electricity spot market. This is consistent with the results of the statistical analysis in Table 5.

Second, the spillovers from all three fossil fuel commodities to electricity markets are highest in the high frequency category, followed by the medium frequency and low frequency categories. This is also consistent with the results of the statistical analysis, shown in Table 6.

On the other hand, the proportion of spillovers from fossil fuel commodities to electricity futures in the medium frequency and low frequency categories are higher than to the electricity spot market in those frequencies. Compared to electricity spot, it appears that the spillovers to electricity futures are higher in the long term. There is an implicit possibility that the speed of information transmission from fossil fuel commodities to electricity futures is slower than to electricity spot.

Finally, it is clear that the patterns of return spillovers from fossil fuel commodities to electricity spot and futures markets are quite different. As we mentioned before, spot prices are determined by renewable infeed and electricity demand while futures prices are determined by fossil fuels and mostly influenced by natural gas. We find that the directional spillover from fossil fuel commodities to the electricity futures market, especially quarterly futures and yearly futures, has some similar patterns with the change of electricity production toward natural gas in Europe. For example, the spillovers to quarterly and yearly futures decreased until 2015 but then tended to increase thereafter. This closely corresponds with the changes in electricity production using natural gas in Europe, which decreased from 2010 to 2015 and has since increased, as indicated by Figure 3.

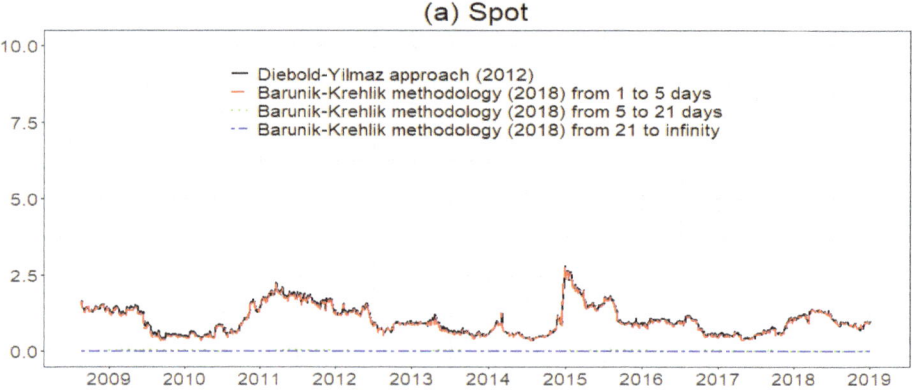

Figure 2. Cont.

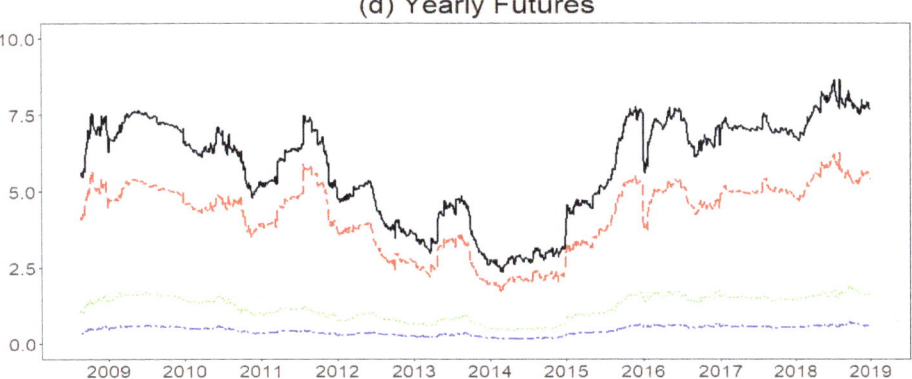

Figure 2. Diebold–Yilmaz approach (2012) directional return spillover (From) and Barunik–Krehlik methodology (2018) directional return spillover (From). Note: spot: EPEX Germany Baseload Spot; monthly futures: EEX Germany Baseload Monthly Futures; quarterly futures: EEX Germany Baseload Quarterly Futures; yearly futures: EEX Germany Baseload Yearly Futures.

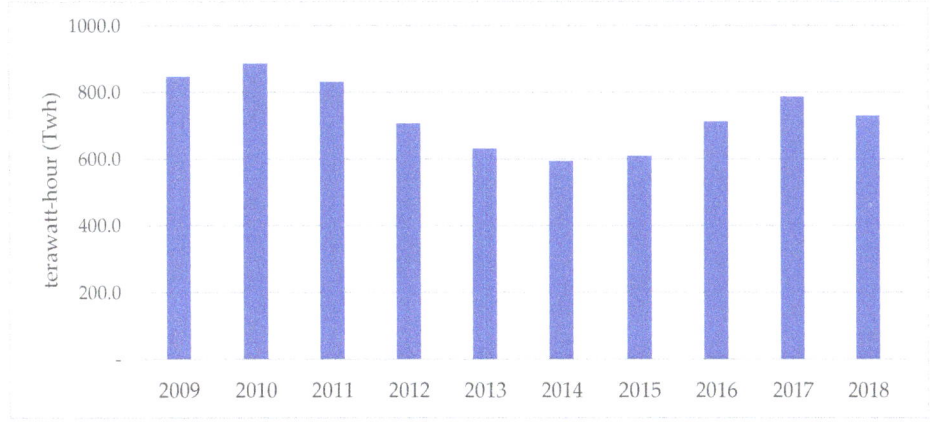

Figure 3. Electricity generation by natural gas in Europe. Data source: BP Statistical Review of World Energy, 2019.

The findings regarding dynamic directional volatility spillover from all fossil fuel commodities to electricity markets are presented in Figure 4 (The pairwise directional volatility spillovers are shown in Figure 2 in the Appendix A). First, it appears that the volatility spillover evolves more fiercely in contrast to return spillovers. In particular, there are many sharp increases, or fluctuations, corresponding to extreme events. This further indicates that volatility spillovers may be more sensitive to extreme events than return spillovers. For example, there are three significant sharp fluctuations in 2009, 2015, and 2016, possibly influenced by the 2008 global financial crisis, the 2014 international oil price's violent shock, and the 2016 rise of coal and natural gas prices. Additionally, the proportions of volatility spillovers from fossil fuel commodities to electricity markets in the high frequency and low frequency categories also increased when the extreme events occurred.

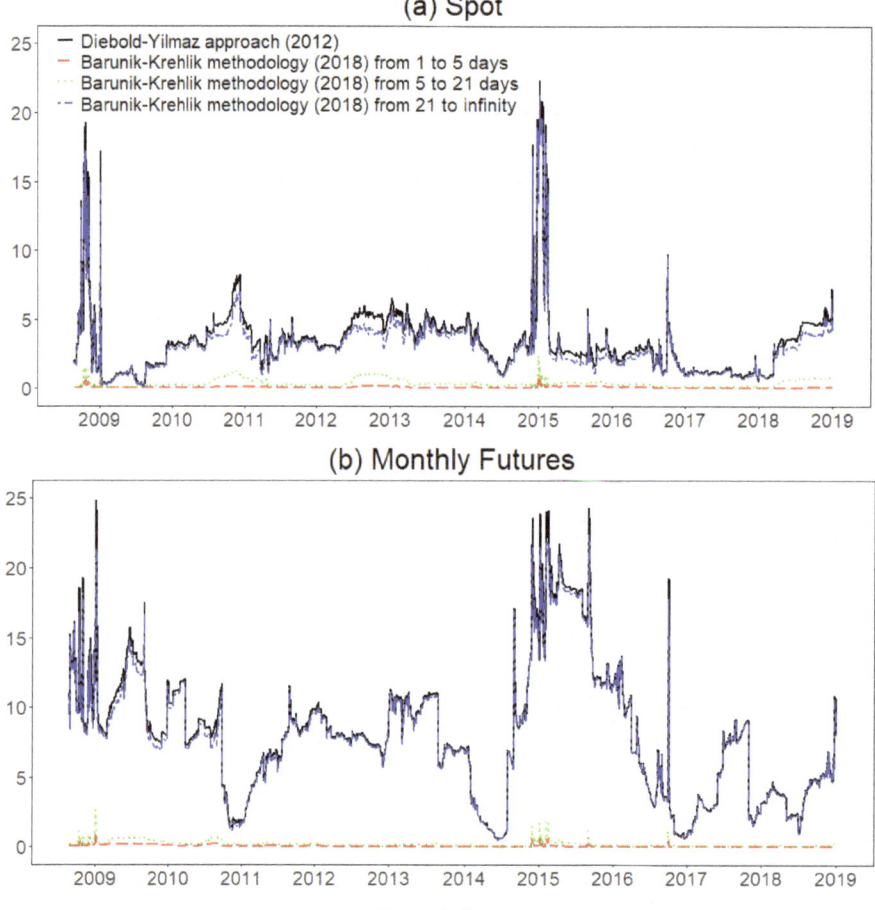

Figure 4. *Cont.*

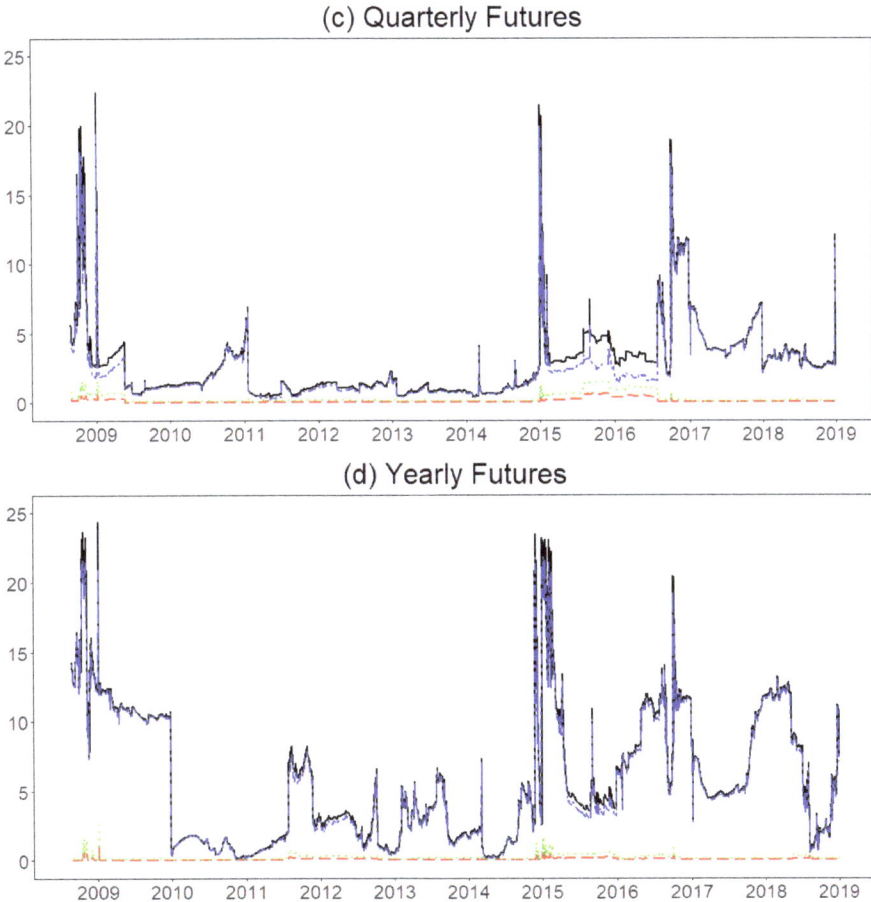

Figure 4. Diebold–Yilmaz approach (2012) directional volatility spillover (From) and Barunik–Krehlik methodology (2018) directional volatility spillover (From). Note: spot: EPEX Germany Baseload Spot; monthly futures: EEX Germany Baseload Monthly Futures; quarterly futures: EEX Germany Baseload Quarterly Futures; yearly futures: EEX Germany Baseload Yearly Futures.

Second, the volatility spillovers from all three fossil fuel commodities to electricity markets are highest in the low frequency category, followed by medium frequency and then high frequency, which is consistent with the results of the statistical analysis in Table 8.

5. Conclusions

This study analyzed return and volatility spillover effects from coal, natural gas, and crude oil fossil fuel markets to electricity spot and futures markets in Europe from 2 January 2007, to 2 January 2019, using a new empirical method in the time-frequency domain frameworks developed by Diebold and Yilmaz [17] and Barunik and Krehlik [18]. The study obtained the following major findings.

First, natural gas has the highest return spillover effect upon the electricity spot market and futures markets, followed by coal and crude oil. The results may be explained by two factors: (1) electricity production favoring one specific fossil fuel and; (2) the different storage cost of fossil fuels.

Second, due to increasing storage costs over time, the return spillovers from fossil fuel commodities to electricity futures with a long delivery period are higher than to electricity futures with a short delivery period. Meanwhile, compared to electricity futures markets, the spot market is more dependent on renewable energy infeed and electricity demand than fossil fuels such as natural gas and coal. Thus, the return spillover effect on electricity futures is higher than on electricity spot markets.

Third, in the frequency domain, we found that the majority of return spillover is generated in the short term which further implies that return shocks are transmitted from fossil fuel markets to electricity markets within only one week.

Contrary to expectations, we found that among the three fossil fuel markets, natural gas has the highest volatility spillover effect on electricity spot markets and monthly and quarterly futures, but not on yearly futures. Similarly, oil exhibits the weakest volatility spillover effect on all electricity markets, also except yearly futures markets.

Additionally, the volatility spillover from natural gas to monthly electricity futures is higher than for electricity futures with a longer delivery period. Meanwhile, crude oil and coal cause the highest volatility spillovers to yearly electricity futures, followed by monthly and then quarterly futures. Except for natural gas, the volatility spillovers from coal and crude oil to electricity futures are higher than in the electricity spot market, which is the same result as that found in the analysis of return spillovers.

In the frequency domain, the majority of volatility spillover is produced in the long term which further indicates that transmitted shocks from fluctuations in fossil fuels have long-lasting effects on electricity market volatility.

Furthermore, we also explored dynamic spillovers by adopting the 400-day moving window method. We found there is a similar pattern in the dynamics of return spillovers from fossil fuel commodities to electricity futures with long delivery periods, and to electricity generation from natural gas. However, we found the dynamic volatility spillovers from fossil fuels to electricity markets are more sensitive to extreme events such as the 2008 global financial crisis, the 2014 international oil price's violent shock, and the 2016 rise in coal and natural gas prices, as shown by volatility spillovers varying sharply when the extreme events occurred.

The results in this paper may be helpful for investors with different investment horizons in Europe to diversify their portfolios, hedge their strategies, and make their risk management plans. For short term investors, constructing well-diversified portfolios consisting of fossil fuels futures, electricity spot and futures is a complicated task, especially in times of financial turmoil. On the other hand, for long-term investors, including the fossil fuels in portfolios composed primarily of electricity spot and futures with different delivery periods could enable them to obtain the long-term diversification benefits.

Author Contributions: Conceptualization, T.N. and S.H.; investigation, T.L. and X.H.; writing—original draft preparation, T.L.; writing—review and editing, X.H., T.N., and S.H.; project administration, S.H.; funding acquisition, S.H. All authors have read and agreed to the published version of the manuscript.

Funding: This work was supported by JSPS KAKENHI Grant Number 17H00983.

Acknowledgments: We are grateful to Keukwan Ryu and two anonymous referees for helpful comments and suggestions.

Conflicts of Interest: The authors declare no conflict of interest.

Appendix A

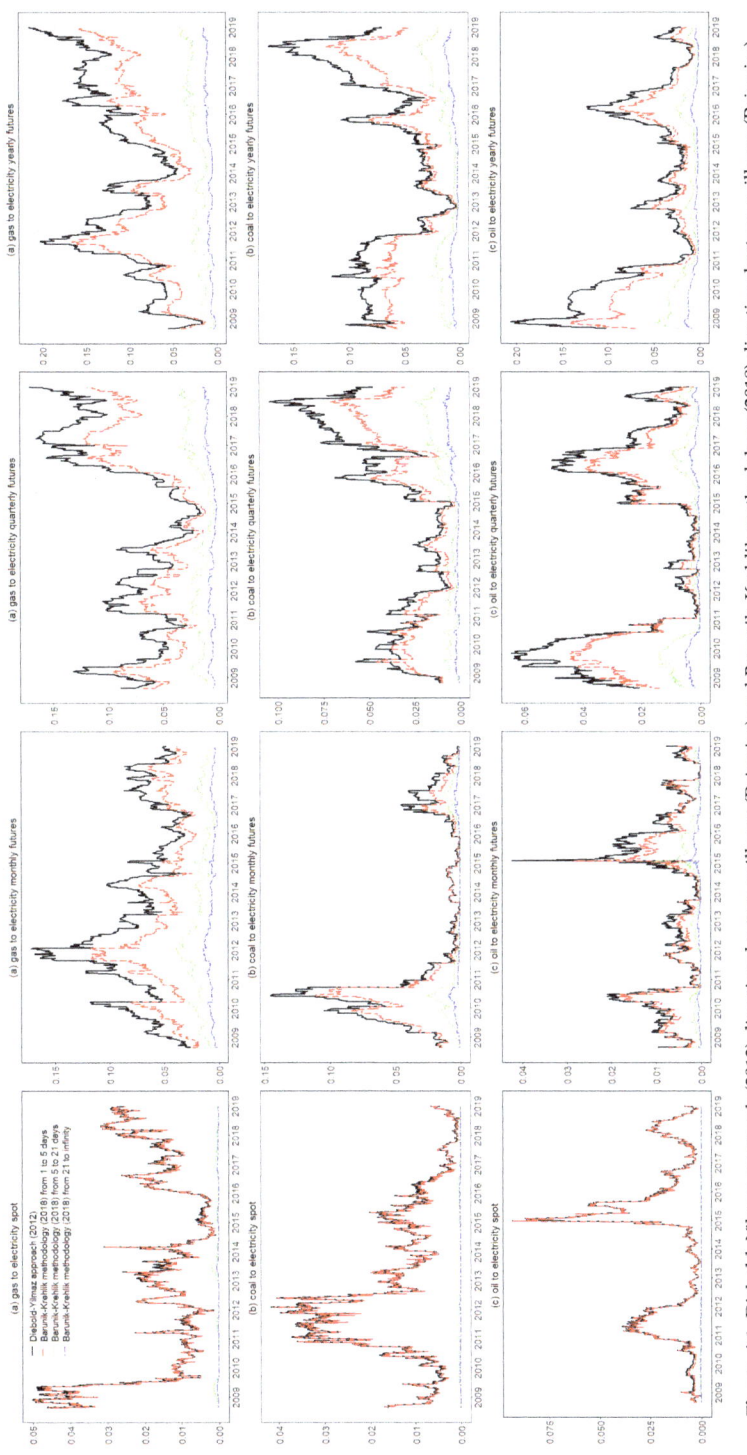

Figure A1. Diebold–Yilmaz approach (2012) directional return spillover (Pairwise) and Barunik–Krehlik methodology (2018) directional return spillover (Pairwise).

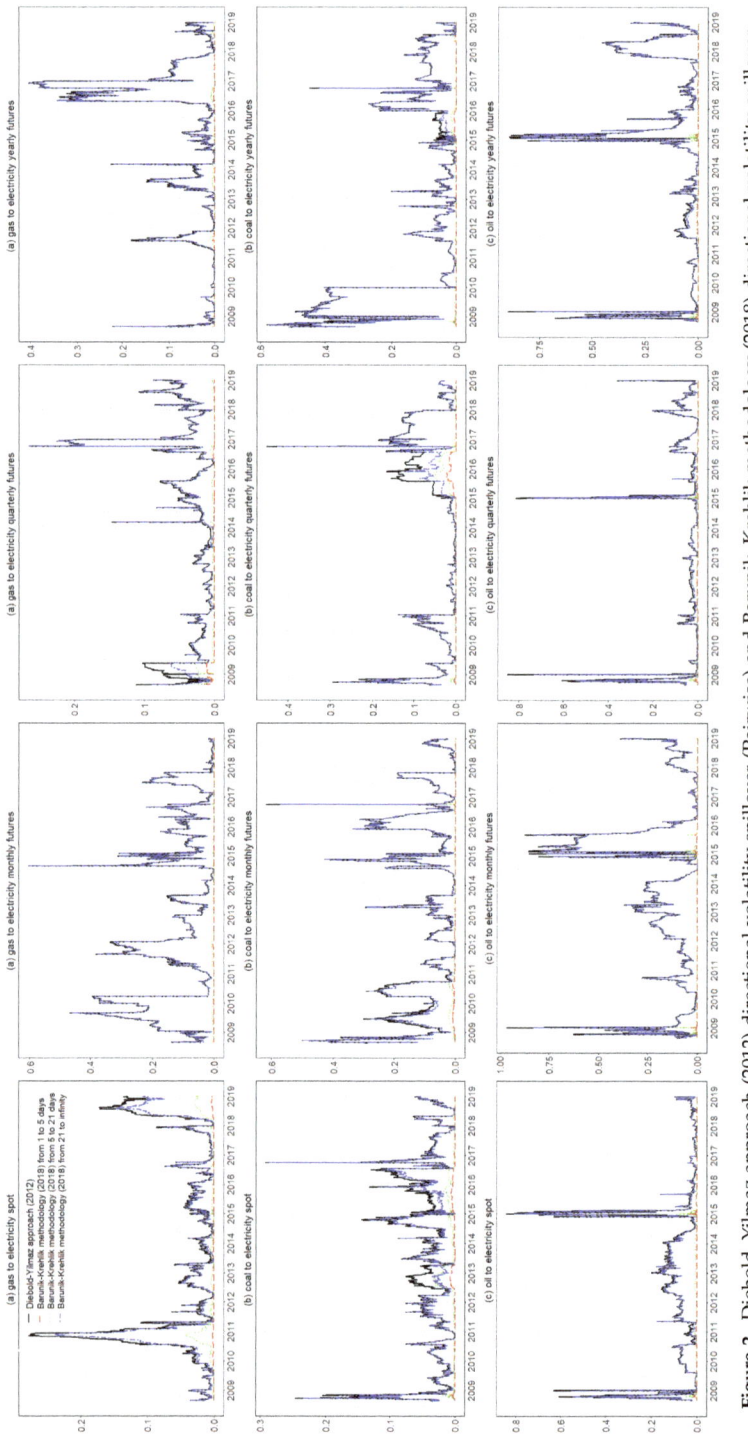

Figure 2. Diebold–Yilmaz approach (2012) directional volatility spillover (Pairwise) and Barunik–Krehlik methodology (2018) directional volatility spillover (Pairwise).

References

1. Bublitz, A.; Keles, D.; Fichtner, W. An analysis of the decline of electricity spot prices in Europe: Who is to blame? *Energy Policy* **2017**, *107*, 323–336. [CrossRef]
2. Kalantzis, F.G.; Milonas, N.T. Analyzing the impact of futures trading on spot price volatility: Evidence from the spot electricity market in France and Germany. *Energy Econ.* **2013**, *36*, 454–463. [CrossRef]
3. Menezes, L.M.; Houllier, M.A.; Tamvakis, M. Time-varying convergence in European electricity spot markets and their association with carbon and fuel prices. *Energy Policy* **2016**, *88*, 613–627. [CrossRef]
4. Paschen, M. Dynamic analysis of the German day-ahead electricity spot market. *Energy Econ.* **2016**, *59*, 118–128. [CrossRef]
5. Huisman, R.; Kilic, M. Electricity futures prices: Indirect storability, expectations, and risk Premiums. *Energy Econ.* **2012**, *34*, 892–898. [CrossRef]
6. Islyaev, S.; Date, P. Electricity futures price models: Calibration and forecasting. *Eur. J. Oper. Res.* **2015**, *247*, 144–154. [CrossRef]
7. Junttila, J.; Myllymäki, V.; Raatikainen, J. Pricing of electricity futures based on locational price differences: The case of Finland. *Energy Econ.* **2018**, *71*, 222–237. [CrossRef]
8. Kallabis, T.; Pape, C.; Weber, C. The plunge in German electricity futures prices Analysis using a parsimonious fundamental model. *Energy Policy* **2016**, *95*, 280–290. [CrossRef]
9. BP Statistical Review of World Energy. Available online: https://www.bp.com/en/global/corporate/energy-economics/statistical-review-of-world-energy.html (accessed on 15 August 2019).
10. Mosquera-López, S.; Nursimulu, A. Drivers of electricity price dynamics: Comparative analysis of spot and futures markets. *Energy Policy* **2019**, *126*, 76–87. [CrossRef]
11. Emery, G.W.; Liu, Q.F. An analysis of the relationship between electricity and natural-gas futures prices. *J. Futures Mark.* **2002**, *22*, 95–122. [CrossRef]
12. Mjelde, J.W.; Bessler, D.A. Market integration among electricity markets and their major fuel source markets. *Energy Econ.* **2009**, *31*, 482–491. [CrossRef]
13. Furió, D.; Chuliá, H. Price and volatility dynamics between electricity and fuel costs: Some evidence for Spain. *Energy Econ.* **2012**, *34*, 2058–2065. [CrossRef]
14. Serletis, A.; Shahmoradi, A. Measuring and Testing Natural Gas and Electricity Markets Volatility: Evidence from Alberta's Deregulated Markets. *Stud. Nonlinear Dyn. Econ.* **2006**, *10*, 1–20. [CrossRef]
15. Green, R.; Larsson, K.; Lunina, V.; Nilsson, B. Cross-commodity news transmission and volatility spillovers in the German energy markets. *Bank. Financ.* **2018**, *95*, 231–243. [CrossRef]
16. Diebold, F.X.; Yilmaz, K. Measuring financial asset return and volatility spillovers, with application to global equity markets. *Econ. J.* **2009**, *119*, 158–171. [CrossRef]
17. Diebold, F.X.; Yilmaz, K. Better to give than to receive: Forecast-based measurement of volatility spillovers. *Int. J. Forecast.* **2012**, *28*, 57–66. [CrossRef]
18. Baruník, J.; Křehlík, T. Measuring the frequency dynamics of financial connectedness and systemic risk. *J. Financ. Econ.* **2018**, *16*, 271–296. [CrossRef]
19. Singh, V.K.; Nishant, S.; Kumar, P. Dynamic and directional network connectedness of crude oil and currencies: Evidence from implied volatility. *Energy Econ.* **2018**, *76*, 48–63. [CrossRef]
20. Kang, S.H.; Lee, J.W. The network connectedness of volatility spillovers across global futures markets. *Phys. A. Stat. Mech. Its Appl.* **2019**, *526*, 120756. [CrossRef]
21. Malik, F.; Umar, Z. Dynamic connectedness of oil price shocks and exchange rates. *Energy Econ.* **2019**. [CrossRef]
22. Lovcha, Y.; Perez-Laborda, A. Dynamic frequency connectedness between oil and natural gas volatilities. *Econ. Model.* **2019**. [CrossRef]
23. Lau, M.C.K.; Vigne, S.A.; Wang, S.X.; Yarovaya, L. Return spillovers between white precious metal ETFs: The role of oil, gold, and global equity. *Int. Rev. Financ. Anal.* **2017**, *52*, 316–332. [CrossRef]
24. Koop, G.; Pesaran, M.H.; Potter, S.M. Impulse response analysis in nonlinear multivariate models. *J. Econ.* **1996**, *74*, 119–147. [CrossRef]
25. Pesaran, H.H.; Shin, Y. Generalized Impulse Response Analysis in Linear Multivariate Models. *Econ. Lett.* **1998**, *58*, 17–29. [CrossRef]
26. Diebold, F.X.; Yilmaz, K. On the Network Topology of Variance Decompositions: Measuring the Connectedness of Financial Firms. *J. Econ.* **2014**, *182*, 119–134. [CrossRef]

27. Toyoshima, Y.; Hamori, S. Measuring the Time-Frequency dynamics of return and volatility connectedness in global crude oil markets. *Energies* **2018**, *11*, 2893. [CrossRef]
28. Fama, E.F.; French, K.R. Commodity future prices: Some evidence on forecast power, premiums, and the theory of storage. *J. Bus.* **1987**, *60*, 55–73. [CrossRef]

© 2020 by the authors. Licensee MDPI, Basel, Switzerland. This article is an open access article distributed under the terms and conditions of the Creative Commons Attribution (CC BY) license (http://creativecommons.org/licenses/by/4.0/).

Article

Examination of the Spillover Effects among Natural Gas and Wholesale Electricity Markets Using Their Futures with Different Maturities and Spot Prices

Tadahiro Nakajima [1,2,*] **and Yuki Toyoshima** [3]

1. The Kansai Electric Power Company Incorporated, Osaka 530-8270, Japan
2. Graduate School of Economics, Kobe University, Kobe 657-8501, Japan
3. Shinsei Bank, Limited, Tokyo 103-8303, Japan; makenaizard@yahoo.co.jp
* Correspondence: nakajima.tadahiro@a4.kepco.co.jp

Received: 22 February 2020; Accepted: 23 March 2020; Published: 25 March 2020

Abstract: This study measures the connectedness of natural gas and electricity spot returns to their futures returns with different maturities. We employ the Henry Hub and the Pennsylvania, New Jersey, and Maryland (PJM) Western Hub Peak as the natural gas price indicator and the wholesale electricity price indicator, respectively. We also use each commodity's spot prices and 12 types of futures prices with one to twelve months maturities and realize results in fourfold. First, we observe mutual spillover effects between natural gas futures returns and learn that the natural gas futures market is integrated. Second, we observe the spillover effects from natural gas futures returns to natural gas spot returns (however, the same is not evident for natural gas spot returns to natural gas futures returns). We find that futures markets have better natural gas price discovery capabilities than spot markets. Third, we observe the spillover effects from natural gas spot returns to electricity spot returns, and the spillover effects from natural gas futures returns to electricity futures returns. We learn that the marginal cost of power generation (natural gas prices) is passed through to electricity prices. Finally, we do not observe any spillover effects amongst electricity futures returns, except for some combinations, and learn that the electricity futures market is not integrated.

Keywords: spillover effect; natural gas; electricity; spot; futures

1. Introduction

The main goal of this study is to clarify differences caused by maturity differences of natural gas and wholesale electricity futures, and differences between their spot and futures due to their commodity characteristics by examining the spillover effects among their futures with various maturities and spot markets. While we can expect the hypothesis of spillover from natural gas market to peak power market because natural gas is often marginal fuel, we can expect the hypothesis that there is no arbitrage trading among their futures and spot markets because we cannot store natural gas and electric energy easily. Moreover, this study might not only clarify the relationships between natural gas futures with one maturity and electricity futures with another maturity, but also reveal market integration by calculating the connectedness indexes among variables.

In the United States of America (USA) as of 2018, the share of electricity generation by fuel is as follows: natural gas is 35.4%, coal 27.9%, nuclear 19.0%, hydroelectric and renewables 16.8%, others 0.9%, according to BP Statistical Review of World Energy 2019, 68th edition [1]. These statistics imply that natural gas is the main fuel for power generation. Moreover, since the cost of procuring natural gas is the marginal cost of power generation, the relationship between the wholesale electricity market and the natural gas market is of great academic and practical interest. Because of this, many previous studies have empirically investigated the relationship between power prices and gas prices from various viewpoints using various methods. This section introduces only previous studies that have examined the North American market, as this study investigates the USA market.

Serletis and Herbert [2] examined the arbitraging mechanisms between the Henry Hub natural gas prices, the Transco Zone 6 (TZ6) natural gas prices, the New York Harbor heating oil, and the Pennsylvania, New Jersey, and Maryland (PJM) power prices. However, these authors showed that the power price series appear to be stationary, whereas both natural gas price series have the unit root. In other words, they argued that the arbitraging mechanism between the price of electricity and the other energies is not effective. Emery and Liu [3] found the cointegrated relationship between natural gas prices series and electricity prices series by examining both the southwestern USA power market and the Pacific Northwest USA power market. Moreover, they noted that natural gas is often the marginal fuel used to generate peak power. Woo et al. [4] investigated whether there is Granger-causality between the natural gas market and the electricity market in both Northern and Southern California. They argued that historical natural gas prices could significantly help predict future electricity prices. Serletis and Shahmoradi [5] examined the relationship between the Alberta gas and power in Canada to indicate the bidirectional causality. Brown and Yücel [6] revealed bidirectional causal nexus between natural gas prices and electricity prices in regional markets by analyzing both the northeastern and the southwestern USA markets. Mjelde and Bessler [7] studied dynamic price information flows among natural gas, uranium, coal, crude oil, and electricity. They showed price information flow from the two regional electricity peak markets to the natural gas market. Mohammadi [8] examined the long-run relation and short-run dynamics between retail electricity prices and three fossil fuel prices using annual data. The estimated long-run equation between electricity, coal, and natural gas prices is statistically significant. The parameters suggest a 1% rise in natural gas prices increases electricity prices by 0.622%.The estimated vector error correction model indicates that natural gas prices Granger cause electricity prices, but not vice versa. Nakajima and Hamori [9] applied Toda and Yamamoto [10], and Cheung and Ng [11] to test the Granger-causal relationships between natural gas prices and electricity prices in the southern USA markets. The results show unidirectional causality in mean from the natural gas market to the electricity market. However, they found no causality in variance. Efimova and Serletis [12] investigated the relationships between fossil fuel and power price volatilities using multivariate generalized autoregressive conditional heteroskedasticity (GARCH) models. Alexopoulos [13] examined the performance of natural gas prices as a predictor for power prices at national and regional levels. They argued that, besides lower gas prices, the growing importance of gas as a predictor for power prices needs the existence of sufficient gas infrastructures and/or competitive market environments. Nakajima [14] examined whether profits can be earned by statistical arbitrage between natural gas futures and electricity futures on the assumption that power prices and natural gas prices have a cointegration relationship. The results of their spark-spread trading simulations show about 30% yield at maximum. However, to our knowledge, no study to date has measured the connectedness of electricity and natural gas spot returns to their futures returns at different maturities.

Diebold and Yilmaz [15] proposed the spillover index. This index can capture how the variables in a system are connected and can assess the shares of forecast error variation in each variable due to shocks of other variables. However, this approach is not the Granger-causality test. In essence, Diebold and Yilmaz's [15] approach only estimates how the variables mutually influence each other, and cannot assess whether the variables are significant enough to predict the other variables. Furthermore, the index was developed based on the vector moving average (VMA) representation of the vector autoregression (VAR) model. Therefore, we must confirm the stationarity of variables in order to adopt this technique.

Numerous studies have adopted Diebold and Yilmaz's [15] approach to analyze spillover effects in a wide variety of markets, not only traditional financial markets, but also commodity markets. However, only literature based on the energy market and published between 2019 and the present is presented henceforth. Yang [16] investigated the connection between international crude oil prices and economic policy uncertainty indexes of developed countries, namely, those of the USA, the United Kingdom (UK), Japan, Germany, France, and Italy. Singh et al. [17] analyzed the volatility spillover connectedness dynamics of crude oil and global asset indicators covering equity, commodities, bonds, and currency pairs. They adopted the West Texas Intermediate (WTI) futures prices as crude oil prices. The stock market indexes are the French CAC, the Japanese NKY, the Chinese SHCOMP, the USA's SPX, and the UK's UKX. Copper, gold, wheat, soybean, and corn are included by the commodities. The bonds include the USA's, the Japanese, the Chinese, the UK's, and the French 10-year indexes. The currency pairs are the EUR-to-USD, the JPY-to-USD, the GBP-to-USD, the CAD-to-USD, and the AUD-to-USD exchange rates. Pham [18] identified the degree and direction of connectedness between various clean energy stock indexes in order to examine whether all clean energy stocks respond homogeneously to crude oil prices. Jin et al. [19] measured the connectedness between the WTI crude oil, Bitcoin, and gold. Wang et al. [20] adopted the Diebold and Yilmaz [15] methodology as one of the approaches necessary to analyze the relationships between electricity, coal, natural gas, and crude oil in the European futures market. Husain et al. [21] calculated the connectedness among crude oil, palladium, titanium, gold, platinum, silver, steel, and the stock index. Albulescu et al. [22] presented the spillover estimates between the WTI crude oil, Australian dollar, Canadian dollar, South African rand, New Zealand dollar, Chilean peso, and the Brazilian real. Chen et al. [23] empirically analyzed the connectedness amongst the whole German power derivative markets. To address the large portfolio of markets in their study, the dynamic network approach can be combined with high-dimensional variable selection techniques. Scarcioffolo and Etienne [24] analyzed the spillover effects of the Northern American natural gas market using daily spot prices from seven locations in the USA and one location in Canada. Kang et al. [25] examined the connectedness among the five agriculture commodity price indexes (meat, dairy, cereals, vegetables, oils, and sugar) and international crude oil using monthly data. Malik and Umar [26] calculated the demand, supply, and risk shocks from crude oil futures prices by using the methodology proposed by Ready [27]. Then, they examined the connectedness among these three shocks, and exchange rates of major oil-exporting and oil-importing countries, namely the Brazilian real, Canadian dollar, Chinese yuan, Indian rupee, Japanese yen, Mexican peso, and Russian ruble. Song et al. [28] examined the dynamics of directional information spillover of returns and volatilities between fossil energy futures, renewable energy industry stocks, and investor sentiment. They selected the crude oil futures, natural gas futures, and coal futures for the fossil energy market. For the renewable energy stock market, they used the Wilder Hill Clean Energy Index, the S&P Global Clean Energy Index, and the European Renewable Energy Index. As for investor sentiment, they utilized the Google search volume index of three keywords related to renewable energy in Google Trends, including "renewable energy," "solar energy," and "wind energy." Xiao et al. [29] studied the relationships among electricity markets of Poland, Germany, France, the Czech Republic, Portugal, Slovakia, Spain, Hungary, Italy, the UK, and northern Europe. Sun et al. [30] explored the price transmission mechanism from international crude oil to sub producer price indexes. They measured the connectedness among the spot prices of Brent crude oil and 15 Chinese sub-producer price indexes. Nakajima and Toyoshima [31] examined spillovers among the North American, European, and Asia–Pacific natural gas markets. He et al. [32] investigated the connectedness between natural gas and BRICS's (Brazil, Russia, India, China, and South Africa) exchange rate. Tiwari et al. [33] analyzed the connectedness between the Food Price Index, the Beverage Price Index, the Industrial Inputs Price Index, the Agricultural Raw Materials Index, the Metals Price Index, and the Fuel Price Index (which comprises crude oil, natural gas, and coal price indexes). Barbaglia et al. [34] studied the volatility spillovers between crude oil, natural gas, gasoline, ethanol, corn, wheat, soybean, sugar, cotton, and coffee futures prices. Guhathakurta et al. [35] studied the connectedness and directional spillover between cocoa, coffee, rubber, soybeans, soya oil, sugar, wheat, palm oil, oats, corn, aluminum, copper, silver, gold, palladium, platinum, and crude oil. Lovcha and Perez-Laborda [36] investigated the dynamic volatility connectedness between the Henry Hub natural gas futures and the WTI crude oil futures traded on the New York Mercantile Exchange (NYMEX). Zhang et al. [37] examined the return and volatility spillover among the natural gas, crude oil, and electricity utility stock indexes in North America and Europe.

This study applies Diebold and Yilmaz's [15] technique to the wholesale electricity market and the natural gas market. We measure the connectedness between power and gas spot returns and futures returns with different maturities. This study adopts the Henry Hub as the natural gas price indicator—one of the most representative natural gas price indexes—and the PJM Western Hub Peak as the wholesale electricity price indicator—which is both the most representative of wholesale power prices and is expected to be the strong link to natural gas prices as marginal cost. This study utilizes each commodity's spot prices and 12 types of futures with one to twelve months maturities. In other words, this study reveals not only the relationship between natural gas prices and wholesale electricity prices, but also the term structure of those prices by measuring the connectedness among these 26 variables.

The calculation results should be able to provide not only novel academic findings, but also useful information for practitioners. Especially, the results may be extremely informative for power generation companies, which hold huge spots and derivatives for both natural gas as raw material, and power as a product. The fluctuations in the price differences between gas and electricity causes concern for most practitioners. Their business risks are complicated because they conclude a variety of procurement and sales contracts, e.g. sales contracts at a fixed price, procurement contracts at prices linked to the other price indexes, and long-term contracts of 10 years or more. It is more efficient to comprehensively evaluate the risks of all contracts that they hold and to hedge their portfolio because it is more expensive to hedge each contract individually. Therefore, it is practical and significant to examine the spillover effects among natural gas and wholesale electricity markets using their futures prices with various different maturities and their spot prices. The relationships between the natural gas and electricity markets in various regions have been studied. However, there is no literature that analyzes the relationships between natural gas and electricity markets using their futures with more than 12 different maturities and their spot prices. Although we can analogize the hypothesis of spillover from the natural gas market to the electricity market from the previous literature, we can expect to discover some differences caused by different maturities of futures, and some differences between their spot and futures due to their commodity characteristics. The main goal of this study is to clarify these differences.

We can expect twofold results by measuring the spillover indexes between these variables. First, we can easily understand at a glance the general characteristics of the portfolio. Diebold and Yilmaz's [15] approach can quantify the degree to which price fluctuations of certain security components of a portfolio have an impact on the portfolio value. Second, we can grasp the market integration among these securities with ease. This approach can present the connectedness not only between any two variables, but also as a whole.

Our contribution to the literature is fourfold. This paper seeks to clarify the relationship and connectedness between natural gas and wholesale electricity markets' spot and futures with one to twelve months maturities. First, we explore the spillover effect of each spot market individually. The spillover effect from natural gas spot returns to electricity spot returns is larger than the spillover effect from gas spot returns to gas futures returns. We argue that the price return of natural gas spot has a greater effect on the price return of wholesale peak power spot, which is directly produced from gas, than on the price return of natural gas futures by arbitrage trading. On the other hand, we find spillover effects from natural gas futures markets to gas spot markets. We estimate that futures markets have better natural gas price discovery capabilities than spot markets. Because we do not observe any evident spillover effects from electricity spot returns to other variables or from all the futures returns to electricity spot returns, we conclude that power spot prices tend to be dependent on power supply and demand at that immediate time due to the limitations on arbitrage trading due to the non-storability of electricity. Second, we examine the spillover effects of natural gas futures markets. The mutual spillover effects between any gas futures are found. We indicate that the natural gas futures market is integrated. Third, we examine the relationships between the futures markets of gas and power.

Our results show spillover effects from natural gas futures returns to power futures returns, which is consistent with previous studies that support Granger causality from natural gas prices to power prices. Conversely, we do not see notable spillover effects from electricity futures returns to natural gas futures returns. Finally, we explore the relationships between power futures markets. We confirm the presence of spillover effects only between the futures returns with maturity differences of six months. Therefore, we conclude that these effects are not caused by arbitrage trading and that the electricity futures market is not integrated. The maximum power load, which is the main determinant of the prices, has almost semi-annual seasonality. Therefore, we observe these spillover effects between six months differences maturities just as a single phenomenon.

The remainder of this paper is organized as follows: Section 2 describes the analyzed data, summary statistics, and preliminary basic analyses; Section 3 explains the adopted methodology; Section 4 presents the empirical results; Section 5 provides a summary of the findings and states the conclusions.

2. Data and Preliminary Analyses

2.1. Data

We adopt the Henry Hub as the natural gas price indicator in our study. Henry Hub is the name of a distribution hub of a natural gas pipeline system in the USA. The price of the natural gas delivered at that hub is the most referenced natural gas price. The Henry Hub futures are listed on the NYMEX.

As our wholesale electricity price indicator, we select the PJM Western Hub Peak. The PJM is one of the regional electricity transmission organizations that coordinates the generation and distribution of electricity in the USA. We use the peak load price series of the western hub in order to capture the relationship between natural gas and power. The PJM Western Hub Peak futures are listed on the NYMEX.

We obtained daily data from January 5, 2009, to December 31, 2018, from Bloomberg. Natural gas prices and electricity prices are cited in the United States dollars per million British thermal units ($/MMBtu) and the United States dollars per megawatt hours ($/MWh), respectively. We obtained each commodity's spot prices and 12 types of futures with one month to twelve months maturities. In other words, the relationships between 26 types of economic variables is analyzed.

Figure 1 shows time plots of natural gas prices. We observe that the fluctuations of all the price series are almost the same. However, only the natural gas spot price spiked in February 2014 and January 2018.

Figure 1. Time series plots of natural gas prices. (n-months-futures signifies futures with n months maturity).

The reason might be that the natural gas spot market temporally became very tight due to a cold temperature wave. Natural gas spots cannot be arbitraged easily with their futures. Therefore, sharp fluctuations in its supply and demand cause sharp price fluctuations.

Figure 2 shows the time plots of electricity prices. The relationship between the spot price and each futures price series is similar to that of the natural gas market. The fluctuations observed in all the price series are almost similar. However, only the spot price spikes frequently. Because the production and consumption of electricity occur concurrently, its supply stability requires simultaneous and equal amounts of power supply and demand. In other words, arbitrage between the spot prices and futures is practically impossible without the use of fuel for power generation. This causes the frequent spikes of the spot prices.

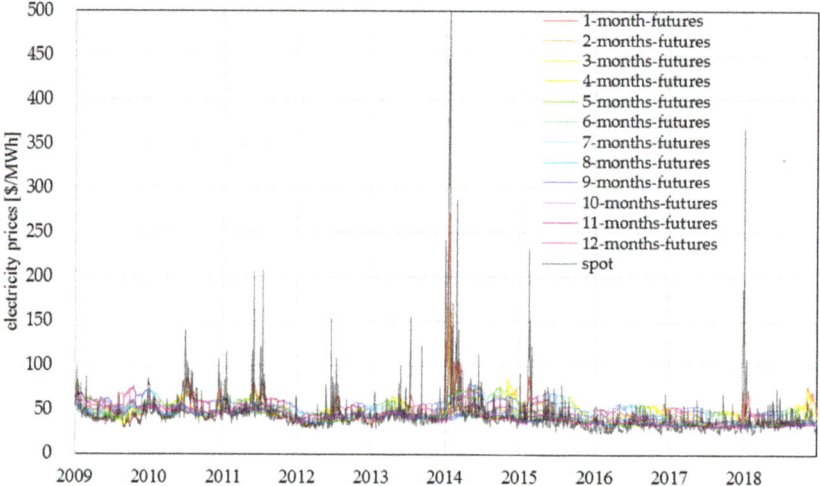

Figure 2. Time series plots of electricity prices. (n-months-futures signifies futures with n months maturity).

Table 1 presents the summary statistics of each return series. There are 2452 observations in each case. The mean of all series are almost zero. All price series have no trend. In other words, loss from holding each security is equal to the risk-free interest cost. In the case of both natural gas and electricity, their spot prices have the outstanding largest maximum return and the outstanding smallest minimum return, compared to their futures. Therefore, only spot prices might fluctuate differently from the futures prices. The standard deviations of natural gas spot returns and electricity spot returns indicate the same. Each sign of skewness has no regularity and is both negative and positive. Most return series are right-skewed, whereas the return series of natural gas futures with four, nine, ten, eleven, and twelve months maturity and electricity futures with one month maturity are left-skewed. We find that all return series are fat-tailed, because their kurtoses have a positive value. Each kurtosis has no relationship with its maturity. The Jarque-Bera statistics calculated from the skewness and kurtosis of each series reject the hypothesis that each series is normally distributed.

2.2. Preliminary Analyses

To calculate Diebold and Yilmaz's [15] index, we need to convert the VAR model to a VMA representation with the condition that all variables remain stationary. Accordingly, we apply the augmented Dickey-Fuller (ADF) unit root test for all series. Table 2 presents the results.

Table 1. Summary statistics of each return series of natural gas and electricity prices.

Return Series	Mean	Maximum	Minimum	Standard Deviation	Skewness	Kurtosis	Jarque–Bera (p-Value)
G0	0.0%	70.1%	−43.9%	4.4%	1.5	45.6	186,350 (0)
G1	0.0%	26.8%	−18.1%	3.1%	0.6	8.3	2983 (0)
G2	0.0%	23.4%	−22.6%	2.8%	0.3	9.3	4143 (0)
G3	0.0%	21.6%	−20.2%	2.6%	0.4	9.7	4730 (0)
G4	0.0%	18.6%	−37.7%	2.5%	−1.0	30.9	79,780 (0)
G5	0.0%	21.7%	−10.1%	2.1%	0.7	9.3	4291 (0)
G6	0.0%	19.3%	−10.9%	2.0%	0.7	10.1	5398 (0)
G7	0.0%	13.7%	−11.3%	1.9%	0.4	8.0	2661 (0)
G8	0.0%	11.6%	−12.8%	1.8%	0.2	7.1	1721 (0)
G9	0.0%	13.2%	−14.8%	1.7%	−0.1	10.0	4954 (0)
G10	0.0%	12.2%	−17.7%	1.6%	−0.6	13.2	10,680 (0)
G11	0.0%	13.0%	−17.2%	1.6%	−0.7	15.4	15,992 (0)
G12	0.0%	11.9%	−18.1%	1.5%	−0.7	16.9	19,984 (0)
E0	0.0%	203.2%	−153.0%	20.2%	0.1	13.8	11,994 (0)
E1	0.0%	55.1%	−67.0%	5.3%	−0.8	42.4	158,546 (0)
E2	0.0%	46.7%	−45.6%	4.0%	0.5	41.0	147,920 (0)
E3	0.0%	37.8%	−30.1%	3.4%	1.9	43.2	166,958 (0)
E4	0.0%	46.1%	−29.5%	3.3%	2.0	57.9	309,913 (0)
E5	0.0%	33.7%	−30.7%	3.1%	1.1	48.1	208,672 (0)
E6	0.0%	40.4%	−32.7%	3.2%	1.6	57.7	307,247 (0)
E7	0.0%	42.9%	−37.5%	3.3%	1.4	63.4	373,262 (0)
E8	0.0%	44.7%	−29.6%	3.2%	2.2	62.4	362,439 (0)
E9	0.0%	40.8%	−27.1%	3.1%	1.8	58.1	311,279 (0)
E10	0.0%	39.8%	−26.6%	3.1%	1.1	57.0	298,650 (0)
E11	0.0%	36.6%	−28.9%	3.0%	0.7	56.3	290,411 (0)
E12	0.0%	33.2%	−32.1%	3.1%	0.6	52.2	247,347 (0)

Note: G0 signifies natural gas spot. Gn signifies natural gas futures with n months maturity. E0 signifies electricity spot. En signifies electricity futures with n months maturity.

Table 2. Unit root tests.

Return Series	Augmented Dickey-Fuller–t Value (p-Value)	
	Exogenous: Constant	Exogenous: Constant, Trend
G0	−16.34 (0.000)	−16.34 (0.000)
G1	−53.70 (0.000)	−53.69 (0.000)
G2	−54.08 (0.000)	−54.08 (0.000)
G3	−53.27 (0.000)	−53.27 (0.000)
G4	−53.59 (0.000)	−53.59 (0.000)
G5	−33.63 (0.000)	−33.62 (0.000)
G6	−29.50 (0.000)	−29.50 (0.000)
G7	−52.53 (0.000)	−52.52 (0.000)
G8	−52.42 (0.000)	−52.41 (0.000)
G9	−52.59 (0.000)	−52.59 (0.000)
G10	−52.50 (0.000)	−52.50 (0.000)
G11	−53.17 (0.000)	−53.18 (0.000)
G12	−30.07 (0.000)	−30.09 (0.000)
E0	−22.95 (0.000)	−22.95 (0.000)
E1	−48.01 (0.000)	−48.00 (0.000)
E2	−21.03 (0.000)	−21.03 (0.000)
E3	−48.29 (0.000)	−48.28 (0.000)
E4	−49.58 (0.000)	−49.57 (0.000)
E5	−49.46 (0.000)	−49.45 (0.000)
E6	−50.06 (0.000)	−50.05 (0.000)
E7	−50.44 (0.000)	−50.43 (0.000)
E8	−49.54 (0.000)	−49.53 (0.000)
E9	−48.79 (0.000)	−48.78 (0.000)
E10	−48.75 (0.000)	−48.74 (0.000)
E11	−49.92 (0.000)	−49.91 (0.000)
E12	−50.38 (0.000)	−50.38 (0.000)

Note: G0 signifies natural gas spot. Gn signifies natural gas futures with n months maturity. E0 signifies electricity spot. En signifies electricity futures with n months maturity.

The ADF test rejects the null hypothesis that all variables are non-stationary. Therefore, we can confirm the availability of the VMA representation.

3. Methodology

We calculate the spillover index proposed by Diebold and Yilmaz [15] in order to capture the relationship between economic variables. This approach can reveal not only pairwise connectedness between any two variables, but also total connectedness between all variables.

We consider the following covariance stationary twenty-six-variable VAR (p):

$$r_t = \sum_{k=1}^{p} \Phi_k r_{t-k} + \varepsilon_t, \tag{1}$$

where r_t is the twenty-six-dimensional vector of return, which must be a stationary series; Φ_k are the 26×26 coefficient matrices; p is the lag length determined by minimizing the Schwarz information criterion; and ε_t is an independently and identically distributed sequence of twenty-six-dimensional random vectors with zero mean and covariance matrix $E(\varepsilon_t \varepsilon_t') = \Sigma$.

We can represent the above covariance stationary twenty-six-variable VAR model in the following VMA:

$$r_t = \sum_{k=0}^{\infty} A_k \varepsilon_{t-k} \tag{2}$$

where $A_k = \sum_{k=1}^{p} \Phi_k A_{t-k}$, A_0 being a 26×26 identity matrix with $A_k = 0$ for $k < 0$.

The spillover effect from the lth to the kth market up to H-step-ahead is defined as the following equation by using the H-step-ahead forecast error variance decompositions:

$$\theta_{kl} = \frac{\sigma_{ll}^{-1} \sum_{h=0}^{H-1} \left(e_k' A_h \Sigma e_l\right)^2}{\sum_{h=0}^{H-1} e_k' A_h \Sigma A_h' e_k} \tag{3}$$

where σ_{ll} is the standard deviation of the error term for the lth equation and e_k is the selection vector, with one as the kth element and zeros otherwise.

Each entry of the variance decomposition matrix is normalized by the row sum, that is, 26, as the pairwise connectedness:

$$\widetilde{\theta_{kl}} = \frac{\theta_{kl}}{\sum_{l=1}^{26} \theta_{kl}} = \frac{\theta_{kl}}{26} \tag{4}$$

The sum of pairwise connectedness is defined as total connectedness:

$$S = \frac{\sum_{k=1}^{26} \sum_{l=1, k \neq l}^{26} \widetilde{\theta_{kl}}}{26} \tag{5}$$

The numerator is the sum of the spillover effects. However, each spillover effect on itself is deducted. In other words, the total connectedness means the sum of the relative proportion of the portfolio's response to a shock.

Moreover, the directional spillover effects received by the kth market from all other markets is measured as:

$$S_{k \cdot} = \frac{\sum_{k=1, k \neq l}^{26} \widetilde{\theta_{kl}}}{26} \tag{6}$$

Similarly, the directional spillover effects transmitted by the kth market to all other markets is measured as:

$$S_{\cdot k} = \frac{\sum_{k=1, k \neq l}^{26} \widetilde{\theta_{lk}}}{26} \tag{7}$$

4. Empirical Results

Table 3 presents the spillover analysis results. This table shows total connectedness, all 676 (= 26 × 26) pairwise connectedness, all 26 directional spillover effects received from all other variables, and all 26 directional spillover effects transmitted to all other variables. The cells in this table are painted gradually from white to dark green, with white representing 0% and dark green representing 10% or more.

Table 3. Spillover index (%).

To \ From	G0	G1	G2	G3	G4	G5	G6	G7	G8	G9	G10	G11	G12	E0	E1	E2	E3	E4	E5	E6	E7	E8	E9	E10	E11	E12	Others
G0	24.8	7.4	6.3	5.7	5.1	5.5	5.3	5.0	4.8	4.5	4.3	4.5	4.4	1.2	0.9	1.4	1.3	1.3	1.1	0.9	0.7	0.6	0.5	0.6	0.9	1.0	2.9
G1	0.7	10.5	9.5	8.5	7.0	7.1	6.9	6.8	6.6	6.1	5.7	5.6	6.0	0.0	0.8	2.2	2.0	1.5	1.0	1.0	1.2	0.9	0.7	0.7	0.5	0.6	3.4
G2	0.6	9.2	10.2	9.4	7.7	7.2	6.7	6.7	6.7	6.4	5.9	5.4	5.5	0.0	0.7	2.1	2.0	1.5	0.9	0.9	1.1	0.9	0.7	0.7	0.5	0.5	3.5
G3	0.6	8.2	9.4	10.1	8.7	7.8	6.7	6.5	6.5	6.5	6.3	5.6	5.1	0.0	0.6	1.7	1.9	1.5	0.9	1.0	1.1	0.9	0.7	0.7	0.5	0.5	3.5
G4	0.6	7.3	8.3	9.4	10.9	8.9	7.2	6.4	6.1	6.3	6.3	5.6	5.0	0.0	0.5	1.6	1.8	1.4	0.9	1.0	1.1	0.9	0.7	0.7	0.5	0.6	3.4
G5	0.5	6.7	7.0	7.6	8.1	10.0	9.0	7.6	6.6	6.2	6.5	6.6	6.2	0.0	0.5	1.4	1.5	1.3	0.9	1.0	1.2	1.0	0.7	0.7	0.6	0.6	3.5
G6	0.6	6.7	6.7	6.8	6.8	9.3	10.4	9.1	7.5	6.3	6.0	6.2	6.5	0.0	0.5	1.4	1.4	1.2	0.9	1.0	1.2	1.0	0.7	0.7	0.6	0.6	3.4
G7	0.5	6.8	6.9	6.7	6.1	8.0	9.2	10.5	9.0	7.3	6.2	5.6	6.1	0.0	0.5	1.5	1.4	1.1	0.9	1.0	1.2	1.0	0.7	0.7	0.6	0.6	3.4
G8	0.5	6.6	7.0	6.9	5.9	7.0	7.7	9.2	10.6	8.9	7.3	5.8	5.4	0.0	0.5	1.5	1.4	1.1	0.9	1.0	1.3	1.0	0.7	0.7	0.6	0.6	3.4
G9	0.5	6.3	6.8	6.9	6.2	6.7	6.6	7.6	9.0	10.8	8.9	7.0	5.6	0.0	0.4	1.5	1.4	1.2	0.9	1.0	1.2	1.0	0.7	0.7	0.6	0.6	3.4
G10	0.5	5.9	6.4	6.8	6.3	7.2	6.4	6.5	7.5	9.0	11.0	8.5	6.8	0.0	0.4	1.4	1.4	1.2	1.0	1.0	1.2	0.9	0.7	0.7	0.6	0.6	3.4
G11	0.5	6.0	6.1	6.3	6.2	7.6	6.9	6.2	6.3	7.4	9.0	11.5	8.8	0.0	0.4	1.4	1.4	1.2	1.0	0.9	1.2	1.0	0.7	0.8	0.6	0.6	3.4
G12	0.6	6.8	6.4	6.0	5.5	7.5	7.6	7.0	6.1	6.3	7.5	9.2	12.1	0.0	0.4	1.5	1.5	1.2	0.9	0.9	1.2	1.0	0.8	0.8	0.6	0.6	3.4
E0	3.4	0.2	0.2	0.2	0.2	0.2	0.2	0.3	0.3	0.3	0.3	0.2	0.2	88.2	3.0	0.6	1.0	0.2	0.3	0.1	0.2	0.1	0.2	0.2	0.2	0.3	0.5
E1	0.6	3.7	3.3	2.9	2.4	2.4	2.5	2.3	2.1	2.0	2.0	1.9	2.7	1.9	49.1	8.9	1.0	0.1	0.2	0.7	5.6	0.8	0.1	0.5	0.0	0.3	2.0
E2	0.0	6.1	5.9	4.6	3.8	3.9	4.0	4.5	4.2	3.7	3.5	3.4	3.6	0.1	4.8	27.3	5.4	0.3	0.0	0.0	1.4	7.8	0.9	0.1	0.5	0.0	2.8
E3	0.3	5.0	5.6	5.2	4.1	3.6	3.2	3.6	4.3	4.0	3.2	3.1	3.1	0.1	0.5	4.8	24.0	4.6	0.2	0.5	0.1	0.8	12.3	2.0	0.6	1.3	2.9
E4	0.5	3.9	4.1	4.6	4.2	3.4	3.2	3.2	3.1	3.9	3.9	3.0	2.9	0.1	0.0	0.2	5.2	26.6	3.6	0.6	0.8	0.5	1.5	14.4	1.6	1.0	2.8
E5	0.6	3.1	3.0	3.1	3.2	3.4	2.8	3.4	3.5	2.8	3.8	3.9	2.5	0.1	0.1	0.1	0.3	4.2	31.8	2.2	0.7	0.9	0.6	1.7	17.1	1.0	2.6
E6	0.3	2.9	3.3	3.2	2.8	3.7	3.4	2.9	3.8	4.1	2.7	3.1	2.9	0.0	0.5	0.1	0.6	0.7	2.1	30.7	3.9	0.7	1.9	0.5	1.4	17.7	2.7
E7	0.1	3.6	3.6	3.8	3.3	4.1	4.8	4.6	3.8	4.3	4.8	3.4	4.1	0.0	3.6	1.7	0.1	1.0	0.7	4.1	32.2	2.2	1.5	2.0	0.2	2.3	2.6
E8	0.1	3.5	3.3	3.0	2.8	3.3	3.9	5.1	4.4	3.2	3.5	4.0	3.0	0.0	0.6	9.9	1.2	0.7	1.0	0.7	2.3	34.0	2.0	2.1	2.1	0.3	2.5
E9	0.1	2.6	3.0	2.8	2.2	2.4	2.2	2.9	4.0	3.6	2.3	2.6	2.7	0.0	0.1	1.1	16.8	1.9	0.6	2.1	1.6	2.0	32.8	2.6	2.4	2.8	2.6
E10	0.5	2.2	2.4	2.7	2.3	2.4	2.4	2.5	2.5	3.5	3.5	2.3	2.7	0.1	0.3	0.1	2.7	17.9	1.8	0.6	2.1	2.0	2.7	33.1	2.8	1.9	2.6
E11	0.9	2.3	1.9	2.0	2.3	2.1	2.0	2.1	2.4	2.1	3.5	3.4	1.8	0.1	0.0	0.7	0.9	2.2	19.3	1.6	0.2	2.2	2.6	3.0	36.0	2.4	2.5
E12	0.6	2.4	2.7	2.2	1.8	2.4	2.2	2.1	2.9	2.8	3.5	2.9	2.6	0.1	0.2	0.1	1.9	1.3	1.1	20.3	2.5	0.3	3.0	2.1	2.4	35.3	2.5
Others	0.6	4.8	5.0	4.9	4.4	4.9	4.7	4.8	4.8	4.7	4.6	4.4	4.1	0.2	0.8	1.9	2.2	2.0	1.7	1.8	1.4	1.2	1.5	1.5	1.5	1.5	75.6

Note: G0 signifies natural gas spot. Gn signifies natural gas futures with n months maturity. E0 signifies electricity spot. En signifies electricity futures with n months maturity.

4.1. Between Spot and Others

We cannot find spillover effects from natural gas spot returns to the other variables except for electricity spot returns (see column G0 of Table 3). The pairwise connectedness from gas spot returns to power spot returns is 3.4% (see row E0 and column G0 in Table 3). We estimate that it is practically difficult to arbitrage gas and electricity futures markets by renting or lending the cash position of natural gas. Due to the quantitative and cost constraints for storing natural gas, natural gas spot prices are more likely to be transmitted to natural gas-fueled peak electricity spot prices than natural gas futures.

The spillover effects from all the gas futures returns to gas spot returns are observed (see row G0 and columns G1 to G12 in Table 3). Although the spot market is susceptible to irregular trading and momentary fluctuations in its supply and demand, the futures markets have enough time to reflect the information related to its price formation, as long as its liquidity is high. Therefore, futures markets have better natural gas price discovery capabilities than spot markets. The pairwise spillover index from electricity futures returns with each maturity to natural gas spot returns is small.

No spillover effects from electricity spot returns to the other variables (see row E0 in Table 3) and from all the futures returns to electricity spot returns occur (see row E0 in Table 3). Electricity spot prices are dependent on the power generation costs and the power market conditions at that time because the non-storability of electricity limits arbitrage trading.

Nakajima [38] and Moutinho et al. [39] examined the relationship between natural gas and wholesale electricity spot prices in Japan and Spain, respectively. Their results reject the hypothesis of the Granger causality between gas and power prices. We cannot argue that the pairwise connectedness indicators between gas and electricity spot returns are consistent with Nakajima [38] and Moutinho et al. [39]. This is assumed to be due to differences between Granger causality and spillover effects and/or regional differences.

4.2. Between Natural Gas Futures

The pairwise spillover indexes between any natural gas futures returns are over 5%, while the spillover effect from each natural gas futures return to itself is about 11% (see rows G1 to G12 and columns G1 to G12 in Table 3). In other words, the natural gas futures market is mostly integrated. Especially, the gas futures markets with a month maturity and with two months maturity are perfectly integrated, because the spillover indexes from these futures returns to themselves are 10.5% (see row G1 and column G1 in Table 3) and 10.2% (see row G2 and column G2 in Table 3), while the indexes from themselves to each other are 9.2% (see row G1 and column G2 in Table 3) and 9.5% (see row G2 and column G1 in Table 3). Furthermore, the spillover effect indexes between futures returns with close maturity differences tend to be greater than those between futures returns with long maturity differences.

4.3. Between Natural Gas Futures and Electricity Futures

We can observe the spillover effects from natural gas futures returns with each maturity to wholesale electricity futures returns with each maturity (see rows E1 to E12 and columns G1 to G12 in Table 3). These calculated results are consistent with previous studies that conclude a Granger causality from natural gas prices to power prices (see Emery and Liu [3] which is one of the most representative previous studies which examines the relationship between natural gas and electricity futures prices). The pairwise spillover indexes from the natural gas futures returns to the electricity futures returns with around the same maturity are larger. The natural gas futures prices with certain maturity affect not only the natural gas futures prices with other maturities, but also the peak power futures prices. The pairwise spillover index from the natural gas futures returns with a one-month period maturity to the wholesale electricity futures returns with a two-month period maturity is 6.1% (see row E2 and column G1 in Table 3).

Conversely, spillover effects from electricity futures to natural gas futures are hardly found (see rows G1 to G12 and columns E1 to E12 in Table 3). The maximum is the pairwise index from the electricity futures returns with a two-month maturity period to natural gas futures returns with a one-month maturity period, which is only 2.2% (see row G1 and column E2 in Table 3). The relatively accurate forecast of electricity demand after two months might affect not only the peak power futures prices but also the fuel futures.

4.4. Between Electricity Futures

We cannot confirm spillover effects between wholesale electricity futures returns, except for the spillover effects between futures returns with maturity differences of six months (see rows E1 to E12 and columns E1 to E12 in Table 3). We should interpret these results by understanding the relationship between the futures price formation and the features of this commodity differently from traditional financial securities.

Futures prices are primarily the expected prices at maturity. However, we should consider arbitrage trading in the spot market. By considering the cost of carrying over the spot position to the maturity of the futures including the opportunity costs and the risk of holding the spot position, we can obtain a non-arbitrage conditional equation for the spot prices and the futures prices. Using this equation, each spot price gives us a unique futures price. However, electricity is a good that is consumed at the same time that it is produced meaning we cannot store electric energy easily. In other words, arbitrage between the electricity spot market and futures markets is practically impossible. Spread trading between futures markets with different maturities is impossible as well, although we cannot deny electricity arbitrage trading through the fuel markets.

On the other hand, the main determinant of electricity prices is the load at that time. Therefore, peak power prices are affected by maximum power loads, which have almost semi-annual seasonality. Table 4 presents the correlation coefficients between the monthly average values of the daily maximum loads. The correlation coefficient between six-month differences is close to one. Moreover, we calculate the correlation coefficients between electricity futures of different maturity to understand the simultaneous relationship as a phenomenon. In Table 5, which shows those coefficients, we observe that the correlation coefficients between six-month differences are relatively large. The cells in this table are painted gradually from white to green, with white representing 0 and dark green representing 1.

Table 4. Correlation coefficients between maximum loads.

Months Difference	1	2	3	4	5	6
Correlation coefficients	0.32	−0.35	−0.88	−0.37	0.38	0.80

Table 5. Correlation coefficients between electricity futures.

Return Series	E1	E2	E3	E4	E5	E6	E7	E8	E9	E10	E11	E12
E1	1.00											
E2	0.43	1.00										
E3	0.05	0.42	1.00									
E4	−0.10	−0.05	0.41	1.00								
E5	−0.01	−0.22	−0.21	0.39	1.00							
E6	0.15	−0.06	−0.33	−0.24	0.32	1.00						
E7	0.40	0.29	−0.15	−0.37	−0.25	0.37	1.00					
E8	0.17	0.63	0.24	−0.24	−0.39	−0.24	0.33	1.00				
E9	−0.10	0.22	0.77	0.25	−0.25	−0.42	−0.27	0.30	1.00			
E10	−0.20	−0.14	0.31	0.79	0.29	−0.23	−0.43	−0.31	0.32	1.00		
E11	−0.10	−0.30	−0.20	0.32	0.79	0.24	−0.23	−0.44	−0.30	0.36	1.00	
E12	0.10	−0.11	−0.39	−0.23	0.29	0.81	0.28	−0.24	−0.45	−0.28	0.32	1.00

Note: En signifies electricity futures with n months maturity.

Therefore, we should interpret these spillover effects between six-month maturity differences just as a seasonal phenomenon. The electricity futures market cannot be considered integrated.

5. Conclusions

We have adopted Diebold and Yilmaz's [15] approach to examine the spillover effects among natural gas and wholesale electricity markets using their futures with different maturities and their spot prices. We used daily data from January 5, 2009, to December 31, 2018, and employed the Henry Hub and the PJM Western Hub Peak as the natural gas price indicator and the wholesale electricity price indicator, respectively. We obtained each commodity's spot prices and 12 types of futures with one to twelve months maturities. In other words, we analyzed the relationships between 26 types of economic variables.

The main results of our analyses are fourfold. First, we find that there are mutual spillover effects amongst natural gas futures returns showing that the natural gas futures market is integrated. The possession of futures with many different maturities has a low diversification effect on a portfolio, although it is effective for directly hedging the spot trading at the applicable maturity. It is reasonable to hold the futures with two-, three-, and four-month period maturities, because these futures largely affect the other variables (see the row Others and columns G2 to G4 in Table 3) and all futures are equally affected by the other variables (see rows G1 to G12 and the column Others in Table 3).

Second, our results show that there are spillover effects from natural gas futures returns to natural gas spot returns, although no spillover effects from natural gas spot returns to natural gas futures returns are evident. From this, we conclude that futures markets have better natural gas price discovery capabilities than spot markets. The natural gas futures markets indicate the prices expected from the long-term economic environment surrounding natural gas. The natural gas spot prices often deviate from their fundamentals because the storage cost is extremely expensive, and the spot prices depend on their fluctuations in supply and demand at that immediate time. Therefore, the futures prices must be utilized when formulating government policies and/or business plans.

Third, we observe spillover effects from natural gas spot returns to electricity spot returns as well as spillover effects from natural gas futures returns to electricity futures returns. The spillover effects from natural gas spot returns to electricity spot returns are larger than the spillover effect from natural gas spot returns to natural gas futures returns. We thus argue that the marginal cost of power generation (natural gas prices) is passed through to electricity prices. The natural gas futures markets are extremely useful in order to control the risks from the electricity markets.

Finally, our results do not show any spillover effects amongst electricity futures returns, except for some combinations, or any spillover effects from electricity futures returns to natural gas futures returns. We confirm the presence of spillover effects between futures with a maturity difference of six months only, concluding these as merely a singular phenomenon. This means it is not caused by arbitrage trading. The maximum power load, which is the main determinant of the prices, has an almost semi-annual seasonality. Thus, we conclude that the electricity futures market is not integrated. We might expect diversification effects by the possession of the futures with many different maturities. However, it is necessary to pay attention to the seasonality in the selection.

Author Contributions: Conceptualization, T.N.; Data curation, T.N.; Formal analysis, T.N.; Validation, Y.T.; Visualization, T.N.; Writing—original draft, T.N.; Writing—review & editing, T.N. All authors have read and agreed to the published version of the manuscript.

Funding: This research received no external funding.

Acknowledgments: The authors would like to thank the anonymous reviewers, whose valuable comments helped improve an earlier version of this paper.

Conflicts of Interest: The authors declare no conflicts of interest.

References

1. BP p.l.c., 1 St James's Square, London SW1Y 4PD, UK. *BP's Statistical Review of World Energy 2019*, 68th ed. Available online: https://www.bp.com/content/dam/bp/business-sites/en/global/corporate/pdfs/energy-economics/statistical-review/bp-stats-review-2019-full-report.pdf (accessed on 4 February 2020).
2. Serletis, A.; Herbert, J. The message in North American energy prices. *Energy Econ.* **1999**, *21*, 471–483. [CrossRef]
3. Emery, G.W.; Liu, W.Q. An analysis of the relationship between electricity and natural-gas futures prices. *J. Futures Mark.* **2002**, *22*, 95–122. [CrossRef]
4. Woo, C.; Olson, A.; Horowitz, I.; Luk, S. Bi-directional causality in California's electricity and natural-gas markets. *Energy Policy* **2006**, *34*, 2060–2070. [CrossRef]
5. Serletis, A.; Shahmoradi, A. Measuring and testing natural gas and electricity markets volatility: Evidence from Alberta's deregulated markets. *Stud. Nonlinear Dyn. Econ.* **2006**, *10*, 10. [CrossRef]
6. Brown, S.P.A.; Yücel, M.K. Deliverability and regional pricing in U.S. natural gas markets. *Energy Econ.* **2008**, *30*, 2441–2453. [CrossRef]
7. Mjelde, J.W.; Bessler, D.A. Market integration among electricity markets and their major fuel source markets. *Energy Econ.* **2009**, *31*, 482–491. [CrossRef]
8. Mohammadi, H. Electricity prices and fuel costs: Long-run relations and short-run dynamics. *Energy Econ.* **2009**, *31*, 503–509. [CrossRef]
9. Nakajima, T.; Hamori, S. Testing causal relationships between wholesale electricity prices and primary energy prices. *Energy Policy* **2013**, *62*, 869–877. [CrossRef]
10. Toda, H.Y.; Yamamoto, T. Statistical inference in vector autoregressions with possibly integrated processes. *J. Econ.* **1995**, *66*, 225–250. [CrossRef]
11. Cheung, Y.W.; Ng, L.K. A causality-in-variance test and its application to financial market prices. *J. Econ.* **1996**, *72*, 33–48. [CrossRef]
12. Efimova, O.; Serletis, A. Energy markets volatility modelling using GARCH. *Energy Econ.* **2014**, *43*, 264–273. [CrossRef]
13. Alexopoulos, T.A. The growing importance of natural gas as a predictor for retail electricity prices in USA. *Energy* **2017**, *137*, 219–233. [CrossRef]
14. Nakajima, T. Expectations for statistical arbitrage in energy futures markets. *J. Risk Financ. Manag.* **2019**, *12*, 14. [CrossRef]
15. Diebold, F.X.; Yilmaz, K. Better to give than to receive: Predictive directional measurement of volatility spillovers. *Int. J. Forecast.* **2012**, *28*, 57–66. [CrossRef]
16. Yang, L. Connectedness of economic policy uncertainty and oil price shocks in a time domain perspective. *Energy Econ.* **2019**, *80*, 219–233. [CrossRef]
17. Singh, V.K.; Kumar, P.; Nishant, S. Feedback spillover dynamics of crude oil and global assets indicators: A system-wide network perspective. *Energy Econ.* **2019**, *80*, 321–335. [CrossRef]
18. Pham, L. Do all clean energy stocks respond homogeneously to oil price? *Energy Econ.* **2019**, *81*, 355–379. [CrossRef]
19. Jin, J.; Yu, J.; Hu, Y.; Shang, Y. Which one is more informative in determining price movements of hedging assets? Evidence from Bitcoin, gold and crude oil markets. *Phys. A* **2019**, *527*, 121121. [CrossRef]
20. Wang, B.; Wei, Y.; Xing, Y.; Ding, W. Multifractal detrended cross-correlation analysis and frequency dynamics of connectedness for energy futures markets. *Phys. A* **2019**, *527*, 121194. [CrossRef]
21. Husain, S.; Tiwari, A.K.; Sohag, K.; Shahbaz, M. Connectedness among crude oil prices, stock index and metal prices: An application of network approach in the USA. *Resour. Policy* **2019**, *62*, 57–65. [CrossRef]
22. Albulescu, C.T.; Demirer, R.; Raheem, I.D.; Tiwari, A.K. Does the U.S. economic policy uncertainty connect financial markets? Evidence from oil and commodity currencies. *Energy Econ.* **2019**, *83*, 375–388. [CrossRef]
23. Chen, S.; Härdle, W.K.; Cabrera, B.L. Regularization approach for network modeling of German power derivative market. *Energy Econ.* **2019**, *83*, 180–196. [CrossRef]
24. Scarcioffolo, A.R.; Etienne, X.L. How connected are the U.S. regional natural gas markets in the post-deregulation era? Evidence from time-varying connectedness analysis. *J. Commod. Mark.* **2019**, *15*, 100076. [CrossRef]

25. Kang, S.H.; Tiwari, A.K.; Albulescu, C.T.; Yoon, S.M. Exploring the time-frequency connectedness and network among crude oil and agriculture commodities V1. *Energy Econ.* **2019**, *84*, 104543. [CrossRef]
26. Malik, F.; Umar, Z. Dynamic connectedness of oil price shocks and exchange rates. *Energy Econ.* **2019**, *84*, 104501. [CrossRef]
27. Ready, R.C. Oil Prices and the Stock Market. *Rev. Financ.* **2018**, *22*, 155–176. [CrossRef]
28. Song, Y.; Ji, Q.; Du, Y.; Geng, J. The dynamic dependence of fossil energy, investor sentiment and renewable energy stock markets. *Energy Econ.* **2019**, *84*, 104564. [CrossRef]
29. Xiao, B.; Yang, Y.; Peng, X.; Fang, L. Measuring the connectedness of European electricity markets using the network topology of variance decompositions. *Phys. A* **2019**, *535*, 122279. [CrossRef]
30. Sun, Q.; An, H.; Gao, X.; Guo, S.; Wang, Z.; Liu, S.; Wen, S. Effects of crude oil shocks on the PPI system based on variance decomposition network analysis. *Energy* **2019**, *189*, 116378. [CrossRef]
31. Nakajima, T.; Toyoshima, Y. Measurement of connectedness and frequency dynamics in global natural gas markets. *Energies* **2019**, *12*, 3927. [CrossRef]
32. He, Y.; Nakajima, T.; Hamori, S. Connectedness between natural gas price and BRICS exchange rates: Evidence from time and frequency domains. *Energies* **2019**, *12*, 3970. [CrossRef]
33. Tiwari, A.K.; Nasreen, S.; Shahbaz, M.; Hammoudeh, S. Time-frequency causality and connectedness between international prices of energy, food, industry, agriculture and metals. *Energy Econ.* **2020**, *85*, 104529. [CrossRef]
34. Barbaglia, L.; Croux, C.; Wilms, I. Volatility spillovers in commodity markets: A large t-vector autoregressive approach. *Energy Econ.* **2020**, *85*, 104555. [CrossRef]
35. Guhathakurta, K.; Dash, S.R.; Maitra, D. Period specific volatility spillover based connectedness between oil and other commodity prices and their portfolio implications. *Energy Econ.* **2020**, *85*, 104566. [CrossRef]
36. Lovcha, Y.; Perez-Laborda, A. Dynamic frequency connectedness between oil and natural gas volatilities. *Econ. Model.* **2020**, *84*, 181–189. [CrossRef]
37. Zhang, W.; He, X.; Nakajima, T.; Hamori, S. How does the spillover among natural gas, crude oil, and electricity utility stocks change over time? Evidence from North America and Europe. *Energies* **2020**, *13*, 727. [CrossRef]
38. Nakajima, T. Inefficient and opaque price formation in the Japan Electric Power Exchange. *Energy Policy* **2013**, *55*, 329–334. [CrossRef]
39. Moutinho, V.; Vieira, J.; Moreira, A.C. The crucial relationship among energy commodity prices: Evidence from the Spanish electricity market. *Energy Policy* **2011**, *39*, 5898–5908. [CrossRef]

© 2020 by the authors. Licensee MDPI, Basel, Switzerland. This article is an open access article distributed under the terms and conditions of the Creative Commons Attribution (CC BY) license (http://creativecommons.org/licenses/by/4.0/).

Article

Can One Reinforce Investments in Renewable Energy Stock Indices with the ESG Index?

Guizhou Liu and Shigeyuki Hamori *

Graduate School of Economics, Kobe University, 2-1, Rokkodai, Nada-Ku, Kobe 657-8501, Japan; liuguizhou0402@gmail.com
* Correspondence: hamori@econ.kobe-u.ac.jp

Received: 8 January 2020; Accepted: 27 February 2020; Published: 4 March 2020

Abstract: Studies on the environmental, social, and governance (ESG) index have become increasingly important since the ESG index offers attractive characteristics, such as environmental friendliness. Scholars and institutional investors are evaluating if investment in the ESG index can positively change current portfolios. It is crucial that institutional investors seek related assets to diversify their investments when such investors create funds in the renewable energy sector, which is highly related to environmental issues. The ESG index has proven to be a good investment choice, but we are not aware of its performance when combined with renewable energy securities. To uncover this nature, we investigate the dependence structure of the ESG index and four renewable energy indices with constant and time-varying copula models and evaluate the potential performance of using different ratios of the ESG index in the portfolio. Criteria such as risk-adjusted return, standard deviation, and conditional value-at-risk (CVaR) show that the ESG index can provide satisfactory results in lowering the potential CVaR and maintaining a high return. A goodness-of-fit test is then used to ensure the results obtained from the copula models.

Keywords: ESG; renewable energy; copula; value-at-risk

---

1. Introduction

In this paper, we seek to illustrate the relationship between the environmental, social, and governance (ESG) index and the renewable energy index. Meanwhile, we want to reveal the potential in combining investments into a portfolio and constructing an index combination that performs better than pre-existing ones. Many institutional investors are focusing on renewable energy investment; these investment portfolios use the renewable energy index as a benchmark, compared to other investment methods, thereby making them much more valuable with a lower exposure to potentially large financial risk. For example, "iShares trust global clean energy ETF" regards "the S&P global clean energy index" as the benchmark, investing 90% of its assets on stocks from the index and others on futures, options, and other contracts. It is also reasonable for large institutional investors to choose stocks or contracts that are full of liquidity and helpful in hedging potential financial risk. For institutional investors, finding a way to reinforce the return yield of their investments in certain fields is the first priority. To reinforce these investments, there should be other types of stocks or bonds contained in the portfolio that are also related to the topic of green energy. After the reinforcement of an investment, the ideal result will be a higher level of returns with a certain risk and a lower level of value at that risk (i.e., the expected shortfall). Of course, diversification will also decrease the potential variation of the investment value. For funds related to the renewable energy index, the ESG index may be a good choice. The ESG index (here, we mainly use the S&P 500 ESG Index (USD) as our research target) is familiar to numerous global investors that are interested in selecting securities with a high standard of sustainability criteria. The ESG refers to environmental, social, and governance factors;

listed companies with these qualifications will normally outperform other companies by providing higher performance in stock prices and bond returns. The ESG index will normally include similar overall sector weights. We want to further investigate if large institutional investors should consider including the ESG index in their investments in the renewable energy sector. There are many selections of ETFs, so we use the index as a substitute for detailed portfolios.

The companies listed in the ESG index are normally attractive investments [1] investigated "green bonds" and companies issuing appealing bonds. Companies have significant incentive to finance themselves by issuing "green bonds". Through these bonds, companies can obtain more attention from the capital market by increasing their ESG score, thereby boosting their stock prices if they are listed. Once the firms are labeled as 'green' and have enough media exposure, they can raise the market demand for their shares. Those companies obtaining greater rewards in environmental management can positively influence their stock price returns [2]. In summary, at the level of company revenue and stock prices, achieving a higher ESG score or being elected into the ESG index is beneficial for the companies by getting them greater attention and more public media exposure.

For institutional investors, such as fund managers holding securities, futures, and forward contracts of environmental interest (in this paper, we mainly focus on the renewable energy index), ESG issues can also be beneficial for portfolio management. There is an association between safeguarding liquidity and hedging [3]; in this paper, we want to prove that the ESG index can be safeguarding tools for renewable energy indices. In previous literatures, the association is proved to exist between strategic cash and lines of credit [4], indicating that the leverage can yield future business opportunities in good times, while the current nonoperational cash can provide guards against cash flow variation. It is also a decent choice for institutional investors to develop the hedging strategies applying futures and forward contracts [5], which are traded actively in fields of renewable energy sector.

The connection between renewable energy and institutional investors are closer than before as the institutional investors keep providing liquidity for the ESG index and renewable energy sector. In many OECD countries, the current situation are low interest rate environment and weak economic growth, in which institutional investors are seeking assets such as renewable energy that provide steady return that have low correlation with other choices of assets [6]. The engagement in the ESG index can help increase potential accounting performance [7]; on the other hand, this index may be successful in reducing possible financial risk, as proven and measured by Hoepner et al. [8], in lower partial moments and value at risk. Meanwhile, investments on the ESG index will gradually motivate the companies in enhancing their business in a more friendly way and help the framing of regulations [9].

To study the combined performance of using multiple indices (corresponding to ETF), we need to illustrate the dependence structure of the ingredients and then examine the detailed performance of possible portfolios. In the field of energy-related securities investment, scholars use Granger casualty [10], the wavelet-based test [11], and copula models [12] to study the co-movements and volatility spillovers among energy stock prices or indices. However, there is still a gap in the research on combining the ESG index and renewable energy securities as a portfolio. Thus, we want to prove the potential benefits of combining the aforementioned indices.

In order to illustrate if investment in the ESG index can help fund managers focus on renewable energy securities, we take the ESG index and the renewable energy stock index as representatives and study the dependence between them. Our research contributes to the literature in the following three dimensions. First, rather than using the VAR model or a Granger casualty test, we use the copula models, which can effectively capture the tail dependence (extreme returns) between the ESG index and the renewable energy stock index, to study the potential static and time varying dependence structures among these variables. Second, based on the estimated marginal distributions and copula parameters, we build four portfolios and discuss in detail whether the portfolio of the ESG index and the renewable energy stock index can be used to reduce potential extreme loss. Risk-adjusted returns, standard deviation in returns, value-at-risk, and conditional value-at-risk (expected shortfall) are used as performance measurements for comparing portfolios and strategies in selecting asset weights. The conditional value-at-risk (CVaR) has proved to be better than value-at-risk (VaR), indicating an overall expected downside risk other than a benchmark [13,14]. It is the same definition mentioned as tail conditional expectation or TailVaR in summary by Artzner et al. [15]. In addition, the CVaR technique can be used after copula model estimation [16,17].

The remainder of our paper is organized as follows. Firstly, we introduce the methodology used in this research in Section 2. The empirical results are provided in Section 3. Section 4 mainly introduces the portfolio performance measurements and comparisons among portfolios. Section 5 is the conclusion.

2. Empirical Methodology

In this research, we first model the marginal distribution for the returns of each financial asset. Then, we use copula models to study the dependence between the ESG index and each of the energy stock indices. Four pairs are thus formed: ESG-NEX, ESG-ECO, ESG-SPGTCED, and ESG-EIRX. Finally, the portfolios of these four pairs will be discussed separately. In this section, we will discuss the marginal distribution models and copula theory. We use not only static copula models but also time varying ones.

2.1. Modeling Marginal Distributions

It is a well-known fact that financial assets have features such as fat tails and serial correlation. Fama supported both fat tailed and skewed distribution [18] in terms of the Mandelbrot stable Paretian distribution, i.e., a stable Levy distribution which was also pointed out in [19]. This can be overcome by using truncated Levy flight (TLF) distribution in [20]. Levy flights appear due to the presence of specific features of autocorrelation [21].

To model the daily returns of the ESG index and energy stock indices, we use the autoregressive moving average (ARMA) model and the standard generalized autoregressive conditional heteroscedasticity (GARCH (1,1)) model [22,23]. The orders of the ARMA and GARCH models are selected by the Bayesian Information Criterion (BIC). We denote the daily returns as the variable Y_t and the shocks as ε_t. The standard residual η_t is an independent and identically distributed variable, having a zero mean and unit variance.

$$Y_t = \phi_{0,y} + \sum_{i=1}^{p} \phi_{i,y} Y_{t-i} + \sum_{j=1}^{q} \theta_{j,y} \varepsilon_{t-j} + \varepsilon_t, \quad \varepsilon_t = \sqrt{h_t}\eta_t, \quad \eta_t \sim iid(0,1) \quad (1)$$

$$h_t = \omega + \beta h_{t-1} + \alpha \varepsilon_{t-1}^2. \quad (2)$$

Before using the copula models to estimate the dependence on standard residuals, we need to transform η_t into a uniform (0,1) distribution. Following Patton [24], we include both parametric (flexible skew t) and nonparametric (empirical distribution function, EDF) models in the transformation of standard residuals. The flexible skew t-model, developed by Hansen [25], has two coefficients, $\nu \in (2, \infty)$ and $\lambda \in (-1, 1)$. This model is preferred since it is flexible in estimating the normal, Student's t, and skew t-distributions with the change of coefficients.

$$\hat{F}(\eta) \equiv \frac{1}{T+1} \sum_{t=1}^{T} 1\{\hat{\eta}_t \leq \eta\} \quad (3)$$

$$\hat{U}(\eta) \equiv F_{skew\ t}(\hat{\eta}_t; \hat{v}, \hat{\lambda}). \quad (4)$$

To ensure we estimate the marginal distribution effectively, several tests are carried out on the results. We use the Ljung–Box test to test if there is an autocorrelation within the (squared) standard residuals up to a chosen lag. The Kolmogorov–Smirnov (KS, Equation (5)) and Cramer Von–Mises (CvM, Equation (6)) tests are chosen for the goodness-of-fit test on the skew t model. We follow the simulation methods in the study of Genest and Rémillard [26] to provide the critical values for KS and CvM. Given the parameters in the marginal distribution models, we simulate S observations and again estimate the models based on the newly produced observations. Then, we compute the KS and CvM statistics and repeat the steps above for T times. The values in the upper $1 - \alpha$ order of the produced T statistics are used as the KS and CvM critical values, separately.

$$KS = \max_n \left| \hat{U}_{(n)} - \frac{n}{N} \right|, \; \hat{U}_{(n)} \text{ is the } n^{\text{th}} \text{ largest value in } \{\hat{U}_{(i)}\}_{i=1}^N \text{ statistics} \quad (5)$$

$$CvM = \sum_{n=1}^N \left(\hat{U}_{(n)} - \frac{n}{N} \right)^2. \quad (6)$$

2.2. Copula Functions

Sklar [27] documented that one can obtain the decomposition of a joint distribution in n dimensions using marginal distributions and the copula function, the latter of which deals with uniform distributed variables. The copula model is a mature method of calculating the level of dependence, including various types of dependence in static and time varying ways. Theoretically, normal (Gaussian) copula can only observe correlations rather than tail dependences. In the Student's t copula model, it is assumed that the tail dependences are symmetric. Models such as the Clayton and rotated Gumbel copula can obtain dependence in the lower tail. Furthermore, if we consider the copula parameters evolving over time, a copula model can be regarded as a time varying model and yield a time-varying level of dependence. In this research, we consider only copula models with two input variables, $Y_{u1,t}$ and $Y_{u2,t}$. Further, we assume that u_1 and u_2 follow uniform distribution. The models used are given in Table 1. In a normal copula, $\phi(\cdot)$ possesses a univariate standard normal distribution, and θ represents a linear correlation coefficient. γ is used to denote the parameter in the Clayton and Gumbel copula [28] model, and dependence is calculated as τ. In this research, we use the rotated Gumbel copula by denoting a new rank in the uniform variables (i.e., we produce $u_1' = 1 - u_1$ and $u_2' = 1 - u_2$). The dependence is then calculated as $\tau = 2 - 2^{-1/\gamma}$. In the Student t copula, the parameter v represents the degrees of freedom, and $t_v^{-1}(\cdot)$ is the inverse of a standard Student's t distribution.

Table 1. Copula functions.

Copula Family	Name
$C(u_1, u_2) = \int_{-\infty}^{\phi^{-1}(u_1)} \int_{-\infty}^{\phi^{-1}(u_2)} \frac{1}{2\pi\sqrt{1-\theta^2}} \exp\left(-\frac{s^2 - 2\theta st + t^2}{2(1-\theta^2)}\right) ds\, dt$	Normal
$C(u_1, u_2) = \left(u_1^{-\gamma} + u_2^{-\gamma} - 1\right)^{-1/\gamma}, \; \gamma \in (0, +\infty), \; \tau = 2^{-1/\gamma}$	Clayton
$C(u_1, u_2) = \exp\{-[(-\ln u_1)^\gamma + (-\ln u_2)^\gamma]^{1/\gamma}\}, \; \gamma \in (1, +\infty), \; \tau = 2^{-1/\gamma}$	Gumbel
$C(u_1, u_2) = \int_{-\infty}^{t_v^{-1}(u_1)} \int_{-\infty}^{t_v^{-1}(u_2)} \frac{1}{2\pi\sqrt{1-\theta^2}} \left(1 + \frac{s^2 - 2\theta st + t^2}{v(1-\theta^2)}\right)^{-\frac{v+2}{v}} ds\, dt$	Student t

Notes: The empirical lower tail dependence of the Clayton copula and rotated Gumbel copula are calculated by $\tau^L = 2^{1/\hat{\gamma}}$ and $\tau^L = 2 - 2^{1/\hat{\gamma}}$. Both tail dependences of the Student's t-copula were calculated by $g_T(\hat{\rho}, \hat{v}) = 2 \times F_{student}\left(-\sqrt{(\hat{v}+1)\frac{\hat{\rho}-1}{\hat{\rho}+1}}, \hat{v}+1\right)$. See Patton [24] for details on the copula functions.

2.3. Generalized Autoregressive Score (GAS) Model in Time Varying Copula

Among the five constant copula models, we select the rotated Gumbel copula and Student's t copula models to further investigate the time varying dependence features. Recently, the generalized autoregressive score (GAS) model, introduced by Creal et al. [29], has been frequently used as a method for obtaining the time varying parameters in copula functions. We denote δ_t as the time-varying copula parameter. Similar to the GARCH model, another term is used to help describe the movement of the autoregressive parameter. The score of the likelihood $I_t^{-1/2} s_t$ is used as the information set. The implicit form of δ_t can be shown as $f_t = h(\delta_t)$. To ensure the movement does not go below 1, the parameter in the rotated Gumbel copula can be written as $\delta_t = 1 + \exp(f_t)$. Similarly, $\delta_t = (1 - \exp\{-f_t\})/(1 + \exp\{-f_t\})$ can be used in Student's t copula to ensure the correlation parameter lies between −1 and 1.

$$f_t = h(\delta_t) \leftrightarrow \delta_t = h^{-1}(f_t) \tag{7}$$

where

$$f_{t+1} = \omega + \beta f_t + \alpha I_t^{-1/2} s_t$$

$$s_t \equiv \frac{\partial}{\partial \delta} \log c(U_{1t}, U_{2t}; \delta_t), \quad I_t \equiv E_{t-1}\left[s_t s_t'\right] = I(\delta_t).$$

3. Empirical Results

3.1. Summary Statistics

We collect data from Bloomberg database, from 28 September, 2007 to 16 April, 2019. In our research, we use the S&P 500 ESG index (the market capitalization-weighted index that measures the performance of securities meeting sustainability criteria) to represent numerous alternative ESG investments. Based on the variable selection of Reboredo [12], we choose four stock indices as our agents for the renewable energy global index.

(1) The Wilder Hill New Energy Global Innovation Index (NEX (for more details on the weights of the NEX index, please refer to https://nexindex.com/whindexes.php.)) is weighted based on globally listed new energy innovation companies and is calculated by Solactive. The index focuses on renewable—solar (27.5% weight), renewable—wind (22.0% weight), energy conversion (5.5% weight), energy efficiency (23.1% weight), energy storage (6.6% weight), renewables—Biofuels % Biomass (8.8% weight), and renewables—others (6.6% weight).

(2) The Wilder Hill Clean Energy Index (ECO (for more details on the weights of the ECO index, please refer to https://wildershares.com/about.php)) is mainly on US-listed clean energy companies and is calculated by the New York Stock Exchange (NYSE). The index focuses on renewable energy supplies (21% weight), energy conversion (21% weight), power delivery and conservation (20% weight), greener utilities (13% weight), energy storage (20% weight), and cleaner fuels (5% weight).

(3) The S&P Global Clean Energy Index (SPGTCED (for more details on the weights of the SPGTCED index, please refer to https://us.spindices.com/indices/equity/sp-global-clean-energy-index)) is weighted based on 30 companies from around the world that are related to clean energy business. The index focuses largely, different from the other three indices, on information technology (24.6% weight). Other weights are allocated on utilities (52.4% weight), industrials (20.8% weight), and energy (2.1% weight).

(4) The European Renewable Energy Total Return Index (ERIX (for more details on the weights of the ERIX index please refer to https://sgi.sgmarkets.com/en/index-details/TICKER:ERIX/)) tracks the stocks of largest European renewable energy companies that are highly involved in wind, water, solar, biofuels, geothermal, and/or marine investments. The index selects the largest companies, in which each component has a minimum weight of 5%. According to the most current weights, the companies are Verbund ag in Austria (21.75% weight), Vestas wind systems a/s in Denmark (20.52% weight), Siemens gamesa renewable ene in Spain (17.67% weight), Edp renovaveis sa in Spain (10.29% weight), and Meyer burger technology in Switzerland (5.95% weight).

Investors cannot directly buy indices but can invest in exchange-traded funds (ETFs), which mirror the indices introduced above. In this paper, we regard each index (or equivalent ETF) as an asset, so they all have prices and returns.

To obtain stationary time series data, we compute the first difference of the natural logarithm of asset prices by multiplying 100 as the returns shown as percentages. Descriptive statistics of the returns are provided in Table 2. Among the four financial assets, only the ESG index gain positive revenues, while the worst case for renewable energy index is negative 0.07% for the average return. In terms of the variation reflected by standard errors and extreme values, the ESG index experiences the lowest volatility. Left-skewed features are found in all assets, indicating a fat tail in the negative returns. The prices and returns of five assets are shown in Figures 1 and 2, respectively. Cointegration and Granger causality between the ESG index and each of other three indices are tested (Table A1).

Table 2. Descriptive statistics of the returns of five financial assets.

	Mean	S.E.	Min	Max	Skewness	Kurtosis	Jarque–Bera
ESG	0.0040	1.0701	−7.2143	8.5659	−0.4448	11.6885	8837
NEX	−0.0378	1.4811	−10.4854	12.0705	−0.4905	11.6112	8702
ECO	−0.0654	2.1128	−14.4673	14.5195	−0.3655	8.2339	3236
SPGTCED	−0.0706	1.9516	−14.9729	18.0927	−0.5195	16.6587	21733
ERIX	−0.0474	2.1309	−16.9652	15.8214	−0.3927	11.2607	7977

Notes: This table summarizes the descriptive statistics of daily returns ranging from 28 September, 2007 to 16 April, 2019 with 2775 observations. ESG, NEX, ECO, SPGTCED, and ERIX are used to represent the S&P 500 ESG index, the Wilder Hill New Energy Global Innovation Index, the Wilder Hill Clean Energy Index, the S&P Global Clean Energy Index, and the European Renewable Energy Total Return Index, separately. The returns of the five assets show negative skewness around zero, large kurtosis, and Jarque–Bera statistics significant at 1%, indicating non-normal distributions.

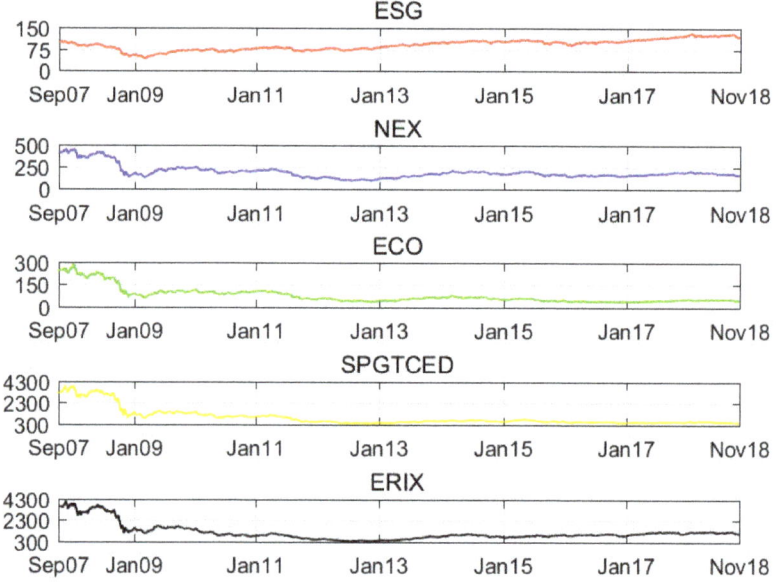

Figure 1. Daily prices of the five assets from 28 September 2007 to 16 April 2019.

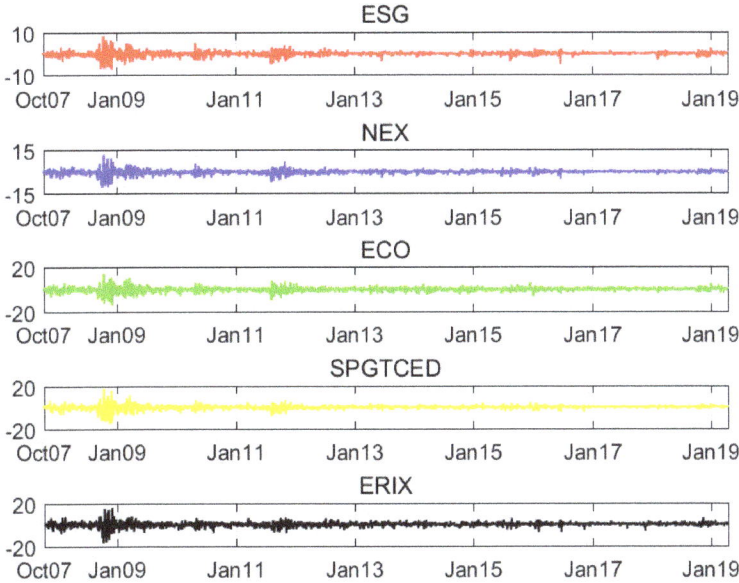

Figure 2. Daily returns of the five assets from 28 September 2007 to 16 April 2019.

3.2. Results for the Marginal Distributions

The results for marginal distributions are shown in Table 3. In order to derive the marginal distributions, the ARMA model is estimated based on stock index returns. Furthermore, the standard GARCH model is estimated on the ARMA residuals to obtain the standard residuals. AR (2) is chosen for all stock indices returns according to the BIC criteria. Almost all coefficients in the standard GARCH model in the five returns are at a 1% significant level. The Ljung–Box test is applied, and the insignificant results document the non-autoregressive features (up to 25 lags) of the standard residuals, as well as the squared ones, at a 10% confidence level. As stated before, the skew t density model is used to model the standard residuals in the standard GARCH model before further calculation in the copula model. The degrees of freedom in the skewed t density model coincide with those from the standard GARCH model estimation. The goodness-of-fit test, including the Kolmogorov–Smirnov (KS) and Cramer Von–Mises (CvM) tests, yields insignificant statistics about the skew t distribution model. The skew t density model appropriately specifies the distribution of the standard residuals.

3.3. Results in the Copula Models

After estimating the AR-GARCH model, the standard residuals for each future return are transformed into a uniform marginal using the skewed t model and empirical distribution function (EDF) methods. Two probability integral transforms are called parametric and nonparametric models, respectively. With these transformed data, we estimate the parameters of various copula models. The constant copula models in this paper include the normal copula, Clayton copula, rotated Gumbel copula, and Student's t copula. The normal copula model yields a correlation between the two series of data. The Clayton copula and rotated Gumbel copula results indicate the dependence of the lower tails between the two series of data. The Student's t copula shows symmetrical dependence structures in the lower and upper tails.

The ESG index and the renewable energy stock indices yield four groups (portfolios) with corresponding dependence levels: 'ESG and NEX', 'ESG and ECO', 'ESG and SPGTCED', and 'ESG and ERIX'. The results of the constant copula models are shown in Table 4. The rankings for the levels of dependence in the four groups are the same in the normal and Clayton copula, with the 'ESG and NEX' group being the highest and the 'ESG and ERIX' group being the lowest. In the results for Student's t copula, the 'ESG and SPGTCED' rather than 'ESG and ECO' group is the second strongest in dependence.

Table 3. Marginal distributions.

	ESG	NEX	ECO	SPGTCED	ERIX
Mean Model					
ϕ_0	0.0547 ***	0.0358 *	0.0138	0.0111	0.0486
S.E.	(0.0129)	(0.0212)	(0.0312)	(0.0249)	(0.0295)
ϕ_1	0.1073 ***	0.1934 ***	0.0501 ***	0.1410 ***	0.0381 **
S.E.	(0.0192)	(0.0193)	(0.0195)	(0.0195)	(0.0192)
ϕ_2	−0.0241	−0.0026	0.0260	0.0068	0.0201
S.E.	(0.0195)	(0.0194)	(0.0195)	(0.0193)	(0.0192)
Variance Model					
ω	0.0070 ***	0.0075 **	0.0383 ***	0.0140 *	0.0316 *
S.E.	(0.0025)	(0.0034)	(0.0136)	(0.0055)	(0.0131)
α	0.1067 ***	0.0738 ***	0.0741 ***	0.0727 ***	0.0721 ***
S.E.	(0.0151)	(0.0116)	(0.0118)	(0.0119)	(0.0125)
β	0.8923 ***	0.9239 ***	0.9160 ***	0.9231 ***	0.9219 ***
S.E.	(0.0139)	(0.0115)	(0.0134)	(0.0120)	(0.0135)
ν	6.2547 ***	8.6623 ***	11.6211 ***	8.2568 ***	6.7693 ***
S.E.	(0.7348)	(1.3224)	(2.3456)	(1.2253)	(0.8643)
Skewed t Density					
λ	6.5835	9.5779	12.6029	8.8251	7.0098
N	−0.1340	−0.1138	−0.1813	−0.1044	−0.0999
Ljung–Box Test					
Q(25)	21.20	14.16	11.30	21.82	23.79
p value	(0.68)	(0.96)	(0.99)	(0.65)	(0.53)
Q^2(25)	25.15	27.32	32.40	18.03	20.23
p value	(0.45)	(0.34)	(0.15)	(0.84)	(0.73)
GoF Tests on Skewed t Distribution Model (p-Value)					
KS	0.30	0.68	0.84	0.17	0.31
CvM	0.34	0.32	0.72	0.15	0.30

Notes: This table provides the coefficients of the AR (2)-GARCH (1,1) model for each return. The standard errors for each parameter are reported in parentheses. ***, **, and * represent significance levels of 1%, 5%, and 10%, respectively. The skewness and degrees of freedom parameters are provided in the skew t model estimation. In the Ljung–Box test, the (Q^2(25)) Q(25) statistics and p-values in parentheses are reported, showing that there is no autocorrelation in (squared) standard residuals up to 25 lags. The results of the goodness-of-fit test (GOF) on the skewed t model are provided as p-values. The null hypothesis of the GOF test (based on 1000 simulations) shows that the data follow the specified skewed t distribution, while the alternative hypothesis is that the data do not follow the specified skewed t distribution. A Cramer Von–Mises (CvM) test is usually regarded as a refinement of the Kolmogorov–Smirnov (KS) test.

Table 4. Constant copula estimations.

	ESG and NEX		ESG and ECO		ESG and SPGTCED		ESG and ERIX	
	Parametric	Semi	Parametric	Semi	Parametric	Semi	Parametric	Semi
Normal Copula								
$\hat{\rho}$	0.7931	0.7933	0.6961	0.6951	0.6887	0.6887	0.6159	0.6159
S.E.	0.0073	0.0075	0.0095	0.0111	0.0094	0.0100	0.0115	0.0127
Log Likelihood	1375.56	1377.16	919.51	915.75	892.20	892.21	661.72	661.68
Clayton Copula								
$\hat{\gamma}$	2.0344	2.0684	1.4834	1.4972	1.4452	1.4601	1.0640	1.0804
S.E.	0.0586	0.0862	0.0633	0.0618	0.0611	0.0591	0.0510	0.0570
$\hat{\tau}^L$	0.7113	0.7153	0.6267	0.6294	0.6190	0.6221	0.5213	0.5265
Log Likelihood	1204.85	1227.97	861.37	865.30	816.19	827.34	564.71	571.87
Rotated Gumbel Copula								
$\hat{\gamma}$	2.3606	2.3754	1.9396	1.9434	1.9357	1.9401	1.6934	1.6984
S.E.	0.0435	0.0472	0.0377	0.0313	0.0441	0.0386	0.0307	0.0314
$\hat{\tau}^L$	0.6587	0.6612	0.5704	0.5714	0.5694	0.5706	0.4942	0.4960
Log Likelihood	1400.45	1415.23	963.50	963.81	941.20	946.61	662.66	666.58
Student's t Copula								
$\hat{\rho}$	0.7986	0.7993	0.6997	0.6996	0.7000	0.7005	0.6228	0.6239
S.E.	0.0085	0.0091	0.0116	0.0122	0.0120	0.0111	0.0149	0.0121
$\hat{\nu}^{-1}$	0.1637	0.1684	0.1340	0.1373	0.1789	0.1810	0.1261	0.1270
S.E.	0.0234	0.0252	0.0192	0.0243	0.0204	0.0240	0.0217	0.0273
$g_T(\hat{\rho}, \hat{\nu})$	0.4014	0.4085	0.2544	0.2597	0.3187	0.3218	0.1838	0.1859
Log Likelihood	1429.05	1432.08	948.61	945.42	954.46	953.46	691.27	691.08

Notes: In this table, the estimated coefficients and the standard errors obtained in the simulation with 100 bootstraps are reported. "Parametric" and "Semi" represent the parametric and semi-parametric models, respectively. The parameter boundaries of the normal copula, Clayton copula, rotated Gumbel copula, Student's t copula, and SJC copula are set as $(-1, 1)$, $(0, \infty)$, $(1, \infty)$, $(-1, 1) \times (2, \infty)$, and $[0, 1) \times [0, 1)$, respectively. In our MATLAB code, the lower bound of the coefficient in the rotated Gumbel copula is set to be 1.1. The log likelihood values are reported using positive signs.

The results for the copula models with time variation are provided in Table 5. Detailed innovations are also presented in Figures 3–6. The GAS model is used in the evolving model for the copula parameter. In four time-varying models, we select the student's t rather than normal copula to better reflect the dependence on both sides. According to Patton [24], rotated Gumbel copula performs better than the Clayton family and they both measure the lower tail dependence; hence, we choose the Gumbel copula in time-varying case. The copula parameter in the next period will be based on the copula parameter in this period and the score of the copula-likelihood.

Table 5. Time-varying copula estimations.

	ESG and NEX		ESG and ECO		ESG and SPGTCED		ESG and ERIX	
	Parametric	Semi	Parametric	Semi	Parametric	Semi	Parametric	Semi
	Rotated Gumbel Copula (GAS)							
$\hat{\omega}$	0.0139	0.0146	−0.0092	−0.0078	−0.0035	−0.0037	−0.0043	−0.0042
S.E.	0.0123	0.0151	0.0152	0.0088	0.0092	0.0115	0.0376	0.0593
$\hat{\alpha}$	0.1364	0.1371	0.1790	0.1720	0.0966	0.1015	0.0545	0.0543
S.E.	0.0329	0.0569	0.0287	0.0528	0.0335	0.0631	0.0329	0.0492
$\hat{\beta}$	0.9522	0.9522	0.8588	0.8750	0.9685	0.9663	0.9899	0.9899
S.E.	0.1367	0.0384	0.1170	0.0253	0.1346	0.0258	0.1413	0.1108
Log Likelihood	1469.21	1485.20	992.11	992.89	983.49	989.66	697.50	702.23
	Student's t Copula (GAS)							
$\hat{\omega}$	0.0480	0.0507	0.1236	0.1212	0.0862	0.0829	0.0109	0.0109
S.E.	0.0219	0.0225	0.0421	0.0373	0.0280	0.0278	0.0132	0.0125
$\hat{\alpha}$	0.0921	0.0984	0.1328	0.1331	0.1354	0.1370	0.0442	0.0442
S.E.	0.0146	0.0146	0.0245	0.0196	0.0184	0.0215	0.0115	0.0092
$\hat{\beta}$	0.9782	0.9770	0.9288	0.9301	0.9501	0.9518	0.9926	0.9926
S.E.	0.0101	0.0103	0.0243	0.0216	0.0157	0.0160	0.0087	0.0081
$\hat{v}^{-1}$	0.1154	0.1281	0.1185	0.1206	0.1467	0.1438	0.1108	0.1108
S.E.	0.0206	0.0249	0.0201	0.0238	0.0223	0.0246	0.0216	0.0213
Log Likelihood	1499.70	1501.32	982.72	979.94	1002.50	1002.36	730.02	730.04

Notes: In this table, the estimated coefficients and standard errors obtained in the simulation with 100 bootstraps are reported. "Parametric" and "Semi" represent the parametric and semi-parametric models, respectively. Log likelihood values are reported using positive signs.

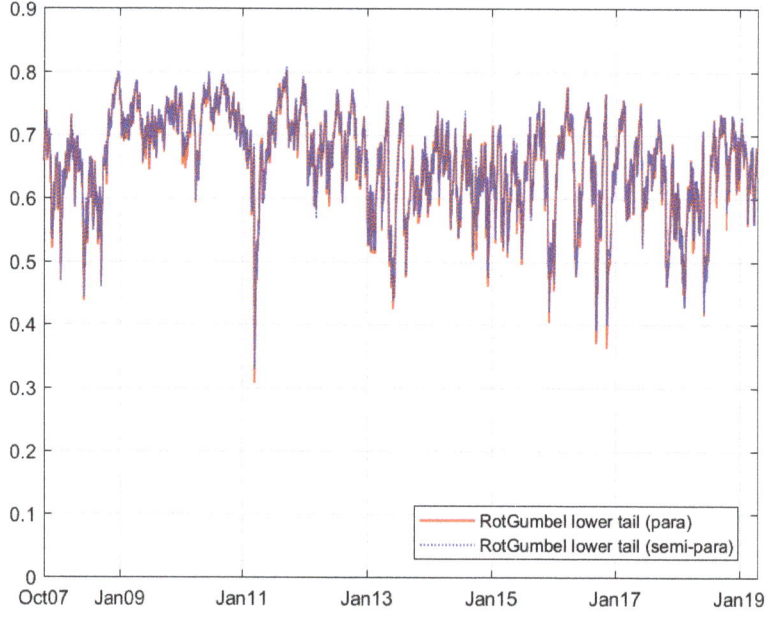

Figure 3. Time varying lower tail dependence between ESG and NEX based on the rotated Gumbel copula (GAS) model.

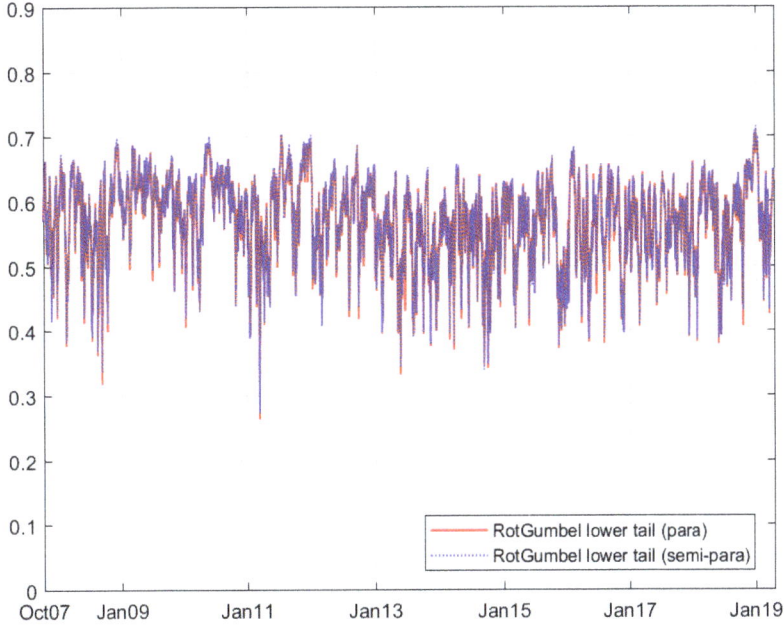

Figure 4. Time varying lower tail dependence between ESG and ECO based on the rotated Gumbel copula (GAS) model.

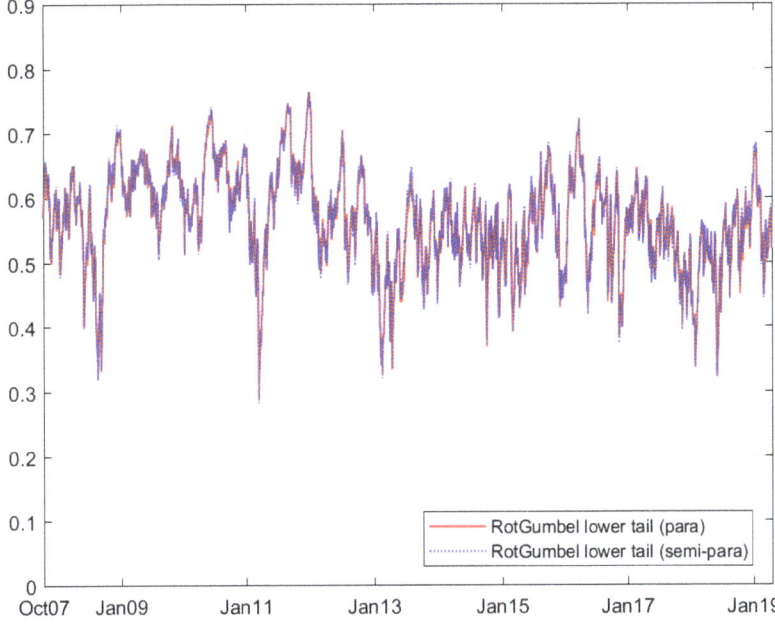

Figure 5. Time varying lower tail dependence between ESG and SPGTCED based on the rotated Gumbel copula (GAS) model.

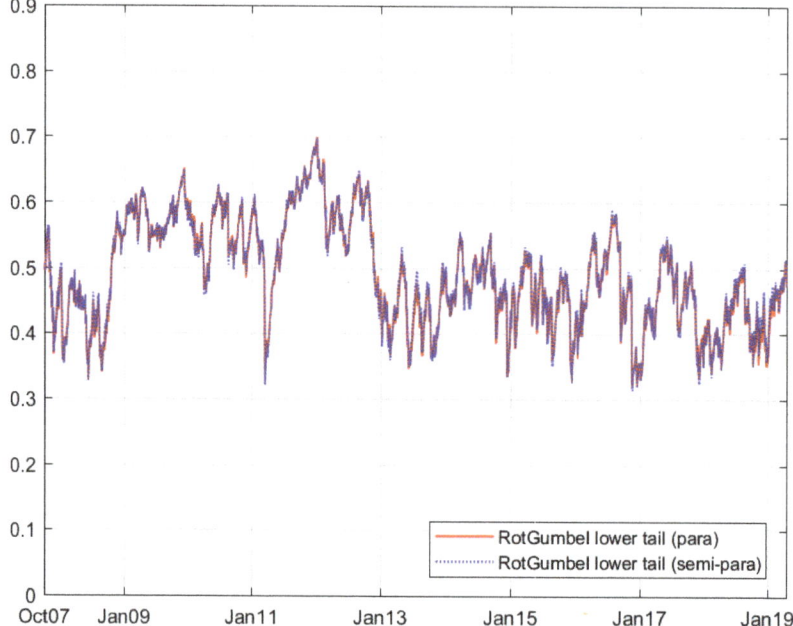

Figure 6. Time varying lower tail dependence between ESG and ERIX based on the rotated Gumbel copula (GAS) model.

3.4. Goodness-of-Fit Test

In order to carry out the goodness-of-fit test for both the constant and time varying copula models, Kolomogorov–Smirnov and Cramer–von Mises methods are used in this paper. For the time varying copula models, the standard residuals should be transformed via the Rosenblatt method. In these two tests, significant statistics indicate that the models based on the data are rejected. The results of the tests on the constant and time varying copula models are reported in Table 6.

In the normal copula models, only the estimations of the parametric method pass the test. For the Clayton copula, no estimation passes the GOF test. For the rotated Gumbel copula, only one semi-parametric case passes the GOF test. For the Student's t copula, only two estimations in semi-parametric cases yielded good results.

In time varying copula models, the p-values in the parametric case are higher than those in the semi-parametric case, although almost all combinations pass both copula models. Comparatively, time varying models under the GAS method offer a better fit for the data than that of the constant copula models.

Table 6. Goodness-of-fit test on the copula models.

	ESG and NEX		ESG and ECO		ESG and SPGTCED		ESG and ERIX	
	KS_C (KS_R)	CvM_C (CvM_R)	KS_C (KS_R)	CvM_C (CvM_R)	KS_C (KS_R)	CvM_C (CvM_R)	KS_C (KS_R)	CvM_C (CvM_R)
			Normal Copula					
Parametric	0.28	0.25	0.23	0.11	0.17	0.17	0.26	0.26
Semi	0.02	0.01	0.01	0.01	0.01	0.01	0.04	0.01
			Clayton Copula					
Parametric	0.01	0.02	0.01	0.04	0.01	0.02	0.02	0.02
Semi	0.01	0.01	0.01	0.01	0.01	0.01	0.01	0.01
			Rotated Gumbel Copula					
Parametric	0.28	0.21	0.58	0.30	0.27	0.13	0.21	0.07
Semi	0.02	0.01	0.16	0.03	0.01	0.01	0.01	0.01
			Student's t Copula					
Parametric	0.26	0.34	0.35	0.09	0.10	0.15	0.36	0.29
Semi	0.03	0.01	0.01	0.01	0.03	0.01	0.13	0.01
			Rotated Gumbel Copula (GAS)					
Parametric	0.01	0.01	0.99	0.99	0.99	0.99	0.01	0.01
Semi	0.42	0.86	0.99	0.99	0.99	0.99	0.54	0.41
			Student's t Copula (GAS)					
Parametric	0.41	0.20	0.47	0.22	0.11	0.08	0.16	0.21
Semi	0.06	0.01	0.01	0.01	0.02	0.01	0.01	0.01

Notes: This table shows the results of the Kolomogorov–Smirnov (KS_C) and Cramer-von Mises (CvM_C) tests on constant copula models of the standard residuals. The KS_R and CvM_R methods test the time varying copula model of the Rosenblatt transform of the standard residuals. P-values that are smaller than 5% are shown in bold font. All results are based on 100 simulations.

4. Portfolio Performance

As shown above, we combine the ESG index with each of the renewable energy stock indices to form pairs and study their dependence structures. We further regard each pair of financial assets as a portfolio, each with different weights. By analyzing the traditional performance standards, such as risk adjusted returns and value-at-risk (VaR), market participators investing in renewable energy index funds and ESG can evaluate the potential revenues and stability behind each portfolio. We denote the portfolios between the ESG index and the renewable energy index as: ESG-NEX, ESG-ECO, ESG-SPGTCED, and ESG-EIRX. In this section, the skew t model and rotated Gumbel copula model will be used to estimate marginal distributions and dynamic dependence. In order to observe the performance of two financial assets, we first need to obtain the linear correlation so we can calculate the covariance and hence the portfolio variance. We simulated the correlation between the two returns using the copula parameters, as shown in Equation (8). Rather than analytically obtaining the results, we used a simulation based on the work by Patton [24].

$$\rho_t \equiv Corr_{t-1}(Y_{u1,t}, Y_{u2,t}) = Corr_{t-1}(\eta_{u1,t}, \eta_{u2,t})$$
$$= E_{t-1}(\eta_{u1,t}, \eta_{u2,t}) = E_{t-1}\left(F_{u1}^{-1}(U_{u1,t}), F_{u2}^{-1}(U_{u2,t})\right) \quad (8)$$

where $u1$ and $u2$ denote different series.

We consider three types of weights for two financial assets. In this research, we denote $w_{1,t}^{(i)}$ as the ith type of weight in the ESG and $w_{2,t}^{(i)} (= 1 - w_{1,t}^{(i)})$ as the weight in the renewable energy stock index at time t. It is assumed that there is no transaction cost, so frequent changes in dynamic weights are allowed and will not affect the returns. The first method involves allocating a constant weight $w_{1,t}^{(1)}$ in the ESG index. We regard the portfolio as a 'static portfolio'. The second method (Equation (9)) is called 'diversified risk-parity', which provides greater capital in assets with less volatility. The third method (Equation (10)) is to consider the method developed by Kroner and Ng [30], which we regard as the 'optimal portfolio'. The latter two dynamic weights are calculated based on conditional variance and covariance.

$$w_{1,t}^{(2)} = \hat{h}_{22,t} / (\hat{h}_{11,t} + \hat{h}_{22,t}) \quad (9)$$

$$w_{1,t}^{(3)} = (\hat{h}_{22,t} - \hat{h}_{12,t}) / (\hat{h}_{11,t} + \hat{h}_{22,t} - 2 \times \hat{h}_{12,t})$$
$$(if\ w_{1,t}^{(3)} < 0,\ w_{1,t}^{(3)} 0 \quad if\ w_{1,t}^{(3)} > 1,\ w_{1,t}^{(3)} 1). \quad (10)$$

The potential performances, including conditional returns, variance, value-at-risk (VaR), and conditional value-at-risk (CVaR, which is also called as expected shortfall (ES)), are obtained based on the following steps. The portfolio contains two assets, denoted as $u1$ and $u2$.

1. We generate dynamic rotated Gumbel copula parameters $\{\hat{\gamma}_t\}_{t=1,2,\dots,T}$ based on the estimated time varying pattern.
2. For time t from 1 to T (total sample size), we generate uniform distributions for two targets $U_{u1,t}$ and $U_{u2,t}$ using $\hat{\gamma}_t$ for S (= 5000) times, and hence the simulated standard residuals and returns based on the estimated marginal distribution parameters. Each element in marginal distributions at time t is stored in a vector of size S × 1.
3. For time t, the portfolio returns are calculated based on each asset return and weight. For asset returns of size S × 1, a lower qth quantile (we selected 1%, indicating a confidence level of 99% for VaR) is regarded as the VaR (Equation (11)). CVaR can also be obtained (Equation (12)).

$$VaR_t^q \equiv F_t^{-1}(q),\ q \in (0,1) \quad (11)$$

$$CVaR_t^q \equiv E[Y_t|F_{t-1},\ Y_t \leq VaR_t^q],\ q \in (0,1). \quad (12)$$

To decide whether the portfolio performs better under a given strategy, we introduce four performance measurements to compare the portfolio and original benchmark returns. Since we investigate the performance of reinforcement after putting the ESG index into the renewable energy index fund, the benchmark portfolio will only include the renewable energy index (i.e., $\omega_{1,t} = 0$, $\omega_{2,t} = 1$). The first measurement is the risk-adjusted returns (RR), indicating returns in units of standard deviation. This comparison requires a standardization of the benchmark (Equation (13)). RR_{com} (RR_{com} is short for 'comparison in RR') obtains the relative change of RR_{w2} in the portfolio compared to RR_{w1} in the benchmark. $RR_{com} > 0$ ($RR_{com} < 0$) yields results allowing the portfolio to acquire a higher (or lower) RR. The second measurement is to compare the standard deviation (SD) of the returns. Similar to RR_{com}, SD_{com} also requires standardization to show the relative changes. $SD_{com} < 0$ ($SD_{com} > 0$) indicates that the portfolio returns are more (less) stable. The third and fourth methods are used to measure the VaR (Equation (15)) and CVaR (ES) (Equation (16)), which are the key elements in portfolio risk management. We compare the portfolio with the benchmark at every time point before taking the average values. VaR_{com} or $CVaR_{com} > 0$ (VaR_{com} or $CVaR_{com} < 0$) demonstrates that the portfolio is more (or less) risky than the benchmark (normally we assume that $VaR_{wj,t}$ and $CVaR_{wj,t}$ ($j = 1$ or 2) have negative signs).

$$RR_{com} = RR_{w2} - RR_{w1} \tag{13}$$

$$SD_{com} = (SD_{w2} - SD_{w1})/SD_{w1} \tag{14}$$

$$VaR_{com} = 1/T \sum_{t=1}^{T} (VaR_{w2,t} - VaR_{w1,t})/VaR_{w1,t} \tag{15}$$

$$CVaR_{com} = 1/T \sum_{t=1}^{T} (CVaR_{w2,t} - CVaR_{w1,t})/CVaR_{w1,t} \tag{16}$$

The results of the comparison between the portfolios under different weights and benchmarks are shown in Table 7. Furthermore, we provide the dynamic CVaR innovation of four different portfolios in Figures 7–10.

The positive values in the comparison of the risk-adjusted returns indicate that the ESG index can increase the profitability level compared to merely investing in the renewable energy index. In the statistics for the returns in the indices listed above, we determine that only the ESG index yields positive revenue. It is thus reasonable that the ESG can increase the revenue in combined investments. Compared to the original status, the combination of ESG and SPGTCED increase the profitability at the highest level, followed closely by the group of ESG and NEX.

When we compare the standard deviation in all groups of investments, we conclude that the ESG index can effectively decrease the deviation of the constructed portfolios. In all forms of strategies, the combination of ESG and ERIX outperform other groups in vanishing volatility. The second-best combination of decreasing the volatility is the group of ESG and ECO. The main reason may be that the original standard deviation values of ERIX and ECO indices are the highest. Therefore, the vanished volatility is obvious.

The ESG index can effectively benefit the portfolios in lowering the value-at-risk level at a large scale. All renewable energy indices can reduce the value-at-risk level by at least 10%, 18%, and 25% under the Naïve static, diversified risk parity, and optimal weights strategies, respectively. In the case of ESG and ERIX, the VaR and CVaR (ES) levels yield the best results.

Thus, the ESG can help ERIX and ECO reduce standard deviation at the largest scale. The ESG and ERIX group is the most successful group in reducing the VaR and CVaR (ES) levels (at nearly 50% of their original levels). These results suggest that using the ESG index in the investment of the renewable energy index should be appealing to most fund managers.

Table 7. Portfolio performance comparison.

Strategy	Measure	ESG and NEX	ESG and ECO	ESG and SPGTCED	ESG and ERIX
Naïve Static	RR_{com}	0.0118	0.0106	**0.0130**	0.0078
	SD_{com}	−0.1672	−0.2871	−0.2640	**−0.2976**
	VaR_{com}	−0.1404	−0.2680	−0.2406	**−0.2851**
	$CVaR_{com}$	−0.1364	−0.2562	−0.2070	**−0.2770**
Diversified risk parity	RR_{com}	0.0167	0.0237	**0.0254**	0.0159
	SD_{com}	−0.2066	−0.4189	−0.3777	**−0.4316**
	VaR_{com}	−0.1852	−0.4244	−0.3379	**−0.4460**
	$CVaR_{com}$	−0.1801	−0.4039	−0.2986	**−0.4305**
Optimal weights	RR_{com}	0.0291	0.0345	**0.0403**	0.0267
	SD_{com}	−0.2722	−0.4890	−0.4502	**−0.4957**
	VaR_{com}	−0.2501	−0.4964	−0.4052	**−0.5084**
	$CVaR_{com}$	−0.2389	−0.4712	−0.3615	**−0.4903**

Notes: In this table, four comparative measurements of the performance between each portfolio and benchmark are reported. For the Naïve Static strategy, we only report the equally (50% to 50%) weighted examples. In each strategy, we report the best results in each row for SD_{com}, VaR_{com}, and ES_{com} in bold font.

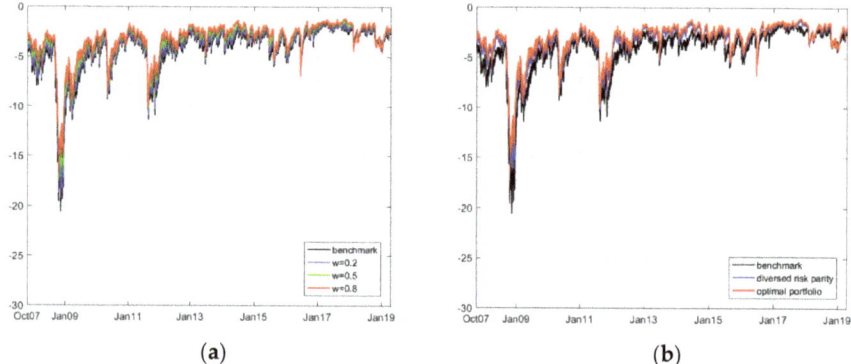

Figure 7. Time varying conditional value-at-risk (CVaR) between the ESG and NEX portfolios ((**a**) is the CVaR under the 'naïve static' strategy; (**b**) is the CVaR under the 'diversified risk parity' and 'optimal weights' strategies).

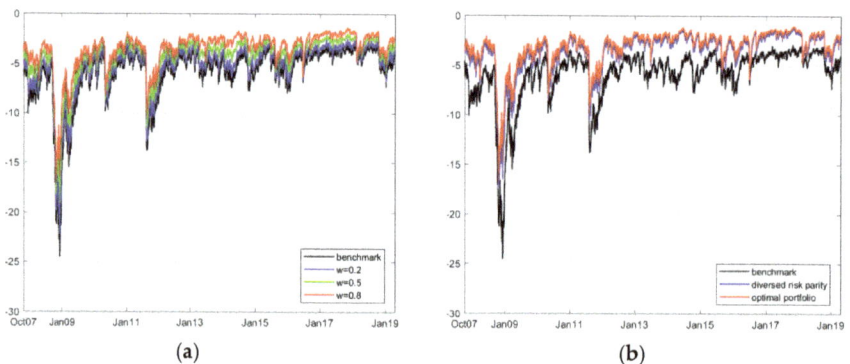

Figure 8. Time varying CVaR between the ESG and ECO portfolios ((**a**) is the CVaR under the 'naïve static' strategy; (**b**) is the CVaR under the 'diversified risk parity' and 'optimal weights' strategies).

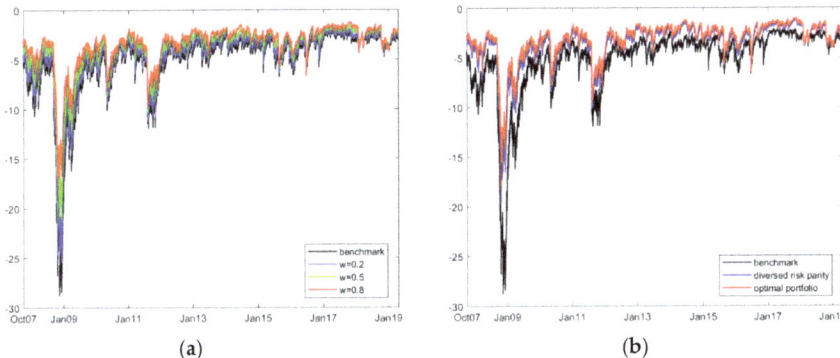

Figure 9. Time varying CVaR between the ESG and SPGTCED portfolios ((**a**) is the CVaR under the 'naïve static' strategy; (**b**) is the CVaR under the 'diversified risk parity' and 'optimal weights' strategies).

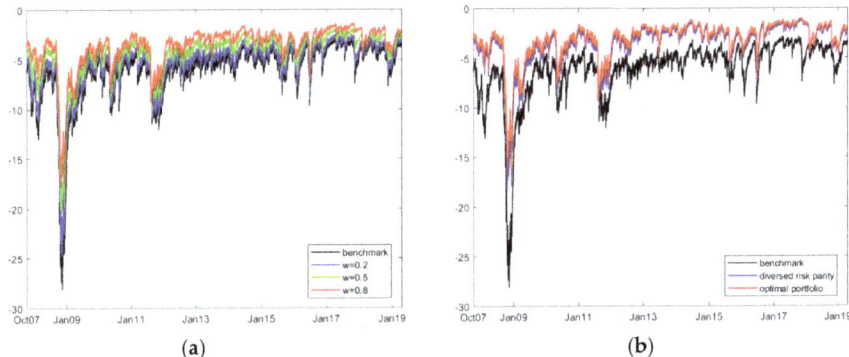

Figure 10. Time varying CVaR between the ESG and ERIX portfolios ((**a**) is the CVaR under the 'naïve static' strategy; (**b**) is the CVaR under the 'diversified risk parity' and 'optimal weights' strategies).

5. Conclusions

The introduction of the ESG index facilitated the diversity of capital investment and fund management. Based on this research, we have studied the essential nature of portfolios and the dependence structure of numerous securities (ETF). The most common and traditional way to reduce the risk and increase the risk-adjusted return is to manage the potential linear correlation among returns. However, to compete in the fierce battlefield of fund management, fund managers need to further investigate the dependence structure in not only constant but also time-varying cases; hence, more precise performance can be simulated in the hypothetical portfolios. Our research, using the constant and time-varying copula models, can help fund managers predict the potential benefits gained from the ESG index if they consider putting the ESG index into the current renewable energy index. These benefits are reasonable since they are similar in topic, and, as stated before, the ESG index has decent historical performance and satisfactory applicability to companies.

Accordingly, we segment the detailed dependence structures into four groups and compare their performances compared with merely investing in the renewable energy index (as the current renewable energy index fund/ETF). Overall, the results suggest that we can trust the ESG index in hedging investment risk and increasing the profitability level in fund management. For example, in the case of investing in the ERIX index, the introduction of the ESG index can effectively reduce the potential value-at-risk level by as much as 50%; meanwhile, this index keeps the simulated revenue positive and more stable. During huge losses, such as the 2008 and 2011 crises, the time varying CVaR graphs reveal that balancing the renewable energy index with the ESG index can be helpful in reducing large losses.

Author Contributions: Conceptualization, S.H.; investigation, G.L.; writing—original draft preparation, G.L.; writing—review and editing, S.H.; project administration, S.H.; funding acquisition, S.H. All authors have read and agreed to the published version of the manuscript.

Funding: This work was supported by JSPS KAKENHI, grant number 17H00983.

Acknowledgments: We are grateful to three anonymous referees for their helpful comments and suggestions.

Conflicts of Interest: The authors declare no conflict of interest.

Appendix A

Table A1. Portfolio cointegration and Granger causality test (*p*-value).

Tests	ESG and NEX	ESG and ECO	ESG and SPGTCED	ESG and ERIX
Cointegration (index)	0.9659	0.8583	0.8763	0.9660
ESG → others (returns)	0.0001	0.0102	0.0702	0.0001
ESG ← others (returns)	0.0001	0.0001	0.0111	0.0001

Notes: In this table, four pairs of index and returns are tested in Engle–Granger cointegration and causality. The maximum lag is set as 5. In the Granger causality test, A → B implies that A is the Granger cause of B, vise verse. We test the cointegration on originally daily index values, while testing the Granger causality on daily returns.

References

1. Tang, D.Y.; Zhang, Y. Do shareholders benefit from green bonds? *J. Corp. Financ.* **2018**. [CrossRef]
2. Klassen, R.; McLaughlin, C. The impact of environmental management on frm performance. *Manag. Sci.* **1996**, *42*, 1199–1214. [CrossRef]
3. Mello, A.S.; Parsons, J.E. Hedging and Liquidity. *Rev. Financ. Stud.* **2000**, *13*, 127–153. [CrossRef]
4. Lins, K.V.; Servaes, H.; Tufano, P. What drives corporate liquidity? An international survey of cash holdings and lines of credit. *J. Financ. Econ.* **2010**, *98*, 160–176. [CrossRef]
5. Shin, Y.J.; Pyo, U. Liquidity Hedging with Futures and Forward Contracts. *Stud. Econ. Financ.* **2019**, *36*, 265–290. [CrossRef]
6. Kaminker, C.; Stewart, F. "The Role of Institutional Investors in Financing Clean Energy OECD Working Papers on Finance, Insurance and Private Pensions, No.23, OECD Publishing." 23. OECD Working Papers on Finance, Insurance and Private Pensions. 2012. Available online: www.oecd.org/daf/fin/wp (accessed on 1 September 2019).
7. Dimson, E.; Karakaş, O.; Li, X. Active ownership. *Rev. Financ. Stud.* **2015**, *28*, 3225–3268. [CrossRef]
8. Hoepner, A.; Oikonomou, I.; Sautner, Z.; Starks, L.; Zhou, X. ESG Shareholder Engagement and Downside Risk. Unpublished work. 2018.
9. Sultana, S.; Zulkifli, N.; Zainal, D. Environmental, Social and Governance (ESG) and Investment Decision in Bangladesh. *Sustainability* **2018**, *10*, 1831. [CrossRef]
10. Keppler, J.H.; Mansanet-Bataller, M. Causalities between CO_2, electricity, and other energy variables during phase I and phase II of the EU ETS. *Energy Policy* **2010**, *38*, 3329–3341. [CrossRef]
11. Reboredo, J.C.; Rivera-Castro, M.A.; Ugolini, A. Wavelet-based test of co-movement and causality between oil and renewable energy stock prices. *Energy Econ.* **2017**, *61*, 241–252. [CrossRef]
12. Reboredo, J.C. Is there dependence and systemic risk between oil and renewable energy stock prices? *Energy Econ.* **2015**, *48*, 32–45. [CrossRef]
13. Rockafellar, R.T.; Uryasev, S. Optimization of Conditional Value-at-Risk. *J. Risk* **2000**, *2*, 21–41. [CrossRef]
14. Pflug, G.C. Some remarks on the value-at-risk and the conditional value-at-risk. In *Probabilistic Constrained Optimization*; Springer: Boston, MA, USA, 2000; pp. 272–281.
15. Artzner, P.; Delbaen, F.; Eber, J.M.; Heath, D. Coherent measures of risk. *Math. Financ.* **1999**, *9*, 203–228. [CrossRef]
16. Deng, L.; Ma, C.; Yang, W. Portfolio optimization via pair copula-GARCH-EVT-CVaR model. *Syst. Eng. Procedia* **2011**, *2*, 171–181. [CrossRef]

17. He, X.; Gong, P. Measuring the coupled risks: A copula-based CVaR model. *J. Comput. Appl. Math.* **2009**, *223*, 1066–1080. [CrossRef]
18. Fama, E.F. The behavior of stock-market prices. *J. Bus.* **1965**, *38*, 34–105. [CrossRef]
19. Hsu, D.A.; Miller, R.B.; Wichern, D.W. On the stable Paretian behavior of stock-market prices. *J. Am. Stat. Assoc.* **1974**, *69*, 108–113. [CrossRef]
20. Mantegna, R.N.; Stanley, H.E. *Introduction to Econophysics: Correlations and Complexity in Finance*; Cambridge University Press: Cambridge, UK, 2000.
21. Figueiredo, A.; Gleria, I.; Matsushita, R.; Da Silva, S. On the origins of truncated Lévy flights. *Phys. Lett. A* **2003**, *315*, 51–60. [CrossRef]
22. Yang, L.; Hamori, S. Dependence structure among international stock markets: A GARCH-copula analysis. *Appl. Financ. Econ.* **2013**, *23*, 1805–1817. [CrossRef]
23. Yang, L.; Hamori, S. Dependence Structure between CEEC-3 and German Government Securities Markets. *J. Int. Financ. Mark. Inst. Money* **2014**, *29*, 109–125. [CrossRef]
24. Patton, A. Copula methods for forecasting multivariate time series. In *Handbook of Economic Forecasting*; Elsevier: Amsterdam, The Netherlands, 2013; Volume 2, pp. 899–960. [CrossRef]
25. Hansen, B.E. Autoregressive conditional density estimation. *Int. Econ. Rev.* **1994**, *35*, 705–730. [CrossRef]
26. Genest, C.; Rémillard, B. Validity of the parametric bootstrap for goodness-of-fit testing in semiparametric models. In *Annales de l'Institut Henri Poincaré, Probabilités et Statistiques*; Institut Henri Poincaré: Paris, France, 2008; Volume 44, pp. 1096–1127.
27. Sklar, M. Fonctions de répartition à n dimensions et leurs marges. *Publ. l'Inst. Stat. l'Univ. Paris* **1959**, *8*, 229–231.
28. Gumbel, E.J. Bivariate exponential distributions. *J. Am. Stat. Assoc.* **1960**, *55*, 698–707. [CrossRef]
29. Creal, D.; Koopman, S.J.; Lucas, A. Generalized autoregressive score models with applications. *J. Appl. Econom.* **2013**, *28*, 777–795. [CrossRef]
30. Kroner, K.F.; Ng, V.K. Modeling asymmetric comovements of asset returns. *Rev. Financ. Stud.* **1998**, *11*, 817–844. [CrossRef]

© 2020 by the authors. Licensee MDPI, Basel, Switzerland. This article is an open access article distributed under the terms and conditions of the Creative Commons Attribution (CC BY) license (http://creativecommons.org/licenses/by/4.0/).

Article

How Does the Spillover among Natural Gas, Crude Oil, and Electricity Utility Stocks Change over Time? Evidence from North America and Europe

Wenting Zhang [1], Xie He [1], Tadahiro Nakajima [1,2] and Shigeyuki Hamori [1,*]

1. Graduate School of Economics, Kobe University, 2-1 Rokkodai, Nada-Ku, Kobe 657-8501, Japan; zhangwenting.kobe@gmail.com (W.Z.); kakyo1515@gmail.com (X.H.); nakajima.tadahiro@a4.kepco.co.jp (T.N.)
2. The Kansai Electric Power Company, Incorporated, 3-chome-6-16 Nakanoshima, Kita-Ku, Osaka 530-8270, Japan
* Correspondence: hamori@econ.kobe-u.ac.jp

Received: 7 January 2020; Accepted: 2 February 2020; Published: 7 February 2020

Abstract: Our study analyzes the return and volatility spillover among the natural gas, crude oil, and electricity utility stock indices in North America and Europe from 4 August 2009 to 16 August 2019. First, in time domain, both total return and volatility spillover are stronger in Europe than in North America. Furthermore, compared to natural gas, crude oil has a greater volatility spillover on the electricity utility stock indices in North America and Europe. Second, in frequency domain, most of the return spillover occurs in the short-term, while most of the volatility spillover occurs over a longer period. Third, the rolling analyses indicate that the return and volatility from 2009 to late 2013 remained stable in North America and Europe, which may be a result of the 2008 global financial crisis, and started to fluctuate after late 2013 due to some extreme events, indicating that extreme events can significantly influence spillover effects. Moreover, investors should monitor current events to diversify their portfolios properly and hedge their risks.

Keywords: natural gas; crude oil; electricity utilities sector index; spillover effect; time–frequency dynamics

1. Introduction

Our paper is aimed at analyzing the return and volatility spillover among natural gas, crude oil, and the electricity utility sector indices across North America and Europe using the methods of Diebold and Yilmaz [1] and Barunik and Krehlik [2] in time and frequency domains. In current commodity markets, energy futures play a major role in economic activities. In particular, as natural gas is cleaner and produces fewer greenhouse emissions than fossil fuels, such as oil and coal, the importance of natural gas has been increasing in the global energy market. Natural gas can be used in many areas, including residential, commercial, industrial, power generation, and vehicle fuels. According to the IEA, natural gas grew 4.6% in 2018, accounting for almost half of the increase in global energy demands (https://www.iea.org/fuels-and-technologies/gas). In both national policy scenarios (gas demand increases by more than a third) and sustainable development scenarios (gas demand will increase slowly by 2030 and return to current levels by 2040), natural gas continues to outperform coal and oil. Meanwhile, crude oil is used to generate electricity, which is also an important raw material for the chemical industry. Thus, we chose the United States (US) and Canada in North America, and Germany, France, the United Kingdom (UK), and Italy in Europe, which are the Group of Seven (G7) member countries, to investigate the spillover among the two energies and electricity utility stocks. According to the BP Statistical Review of World Energy 2019 [3], the consumption shares of natural gas and crude oil in 2018 were as follows: US (16.6%), Canada (2.5%), Germany (2.3%), France (1.7%), the UK (1.4%), and Italy (1.1%). These six countries are major consumers of natural gas and crude oil.

With the development of financial globalization, the financial community is paying increasingly more attention to the transmission of dynamic return links and volatility throughout the capital market. In the situation of a market crash or crisis, portfolio managers and policymakers need to take some actions to prevent the risk of transmission. Therefore, empirical research on the intensity of spillovers provides insights for accurate predictions of returns and volatility. In particular, investors should know, especially in recent years, how fluctuations in natural gas, crude oil, and electricity utilities stock indices affect the risk and value of their investment portfolios. In addition, from 2009 to 2019, several major events influenced these two energy markets and the stock market. These extreme events led to fluctuations in the return and volatility spillover among the three markets in North America and Europe. Hence, understanding the return and volatility spillover caused by financial shocks is not only essential for investors in terms of risk management and portfolio diversification but also for policymakers in developing appropriate policies to avoid impacts from future extreme events. Tian and Hamori [4] have also indicated that policymakers need to understand the transmission mechanism of volatility shock spillover that leads to financial instability.

The main contributions of our study can be summarized as follows. First, as far as we know, this is the first study to investigate the return and volatility spillover among natural gas, crude oil, and the electricity utility sector indices in North America and Europe, respectively, using the Diebold and Yilmaz [1] method for time domain and the Barunik and Krehlik [2] method for frequency domain. Second, we separately analyzed the return and volatility spillover in North America and Europe between the two energy futures and electricity utility stocks to determine the similarities and differences of the spillover effects in the two regions. Third, we employ a rolling analysis to examine the dynamics of the connectedness of return and volatility in time and frequency domains.

The remainder of our paper is described as follows. Section 2 provides a literature review. Section 3 describes the empirical techniques. In Section 4, we explain the data and the descriptive statistics through a preliminary analysis. In Section 5, we report the empirical results of the spillover effects and the moving window analysis. Finally, we conclude our analysis in Section 6.

2. Literature Review

There are numerous studies in the literature investigating spillover effects on the relationship between crude oil and stock markets. Arouri et al. [5] used the generalized vector autoregression (VAR)–generalized autoregressive conditional heteroskedasticity (GARCH) model to analyze the volatility spillover between oil and the stock markets in Europe and the US. Using the sector data, they found that oil and sector stock returns have significant volatility spillover. Soytas and Oran [6] used the Cheung–Ng approach (Cheung and Ng [7]) to analyze the volatility spillover between the world oil market and electricity stock returns in Turkey. They found new information that was not found through conventional causality tests using aggregated market indices. Arouri et al. [8] used a recent generalized VAR–GARCH model to investigate the return and volatility spillover between the oil and stock markets among Gulf Cooperation Council (GCC) countries from 2005 to 2010. They found that the return and volatility spillover between them was significant enough for investors to diversify their portfolios. Nazlioglus et al. [9] investigated the volatility transmission between crude oil and some agricultural commodity markets (wheat, corn, soybean, and sugar). They found that the good price crisis and risk transmission has significantly affected the dynamics of volatility spillover. Nakajima and Hamori [10] analyzed the relationship among electricity prices, crude oil prices, and exchange rates. They found that exchange rates and crude oil Granger cause electricity prices neither in mean nor in variance.

Despite the many well-documented studies on the spillover between crude oil and the stock market, there are relatively few studies on the natural gas and financial markets. Ewing et al. [11] analyzed the volatility spillover between oil and natural gas markets using the GARCH model.

Acaravci et al. [12] investigated the long-term relationship between natural gas prices and stock prices using the vector error correction model developed by Johansen and Juselius [13].

Diebold and Yilmaz [1,14,15] developed the methodology of analyzing the connectedness in time domain based on the variance decomposition of the forecast error to assess the share of forecast error variation in its magnitude and direction; Barunik and Krehlik [2] then extended this connectedness to frequency domain to show the spillover effect from different frequency ranges. Many researchers have applied these empirical techniques to investigate the connectedness between markets in time domain or both in time and frequency domains.

Maghyereh et al. [16] analyzed the connectedness between oil and equities in 11 major stock exchanges in time domain. They found a robust transmission from the crude oil market to the equity market, which grew stronger from mid-2009 to mid-2012. Duncan and Kabundi [17] investigated the domestic and foreign sources of volatility spillover in South Africa in time domain. In addition, Liow [18] characterized the conditional volatility spillover among G7 countries in regard to public real estate, stocks, bonds, money, and currency, both domestically and internationally in time domain. Sugimoto et al. [19] examined the spillover effects on African stock markets during the global financial crisis and the European sovereign debt crisis in time domain.

Toyoshima and Hamori [20] researched the connectedness of return and volatility in the global crude oil markets in time and frequency domains. They found that the Asian currency crisis (1997–1998) and the global financial crisis (2007–2008) generated an increase in return and volatility spillover effects. Lovcha et al. [21] characterized the dynamic connectedness between oil and natural gas volatility in frequency domain. Ferrer et al. [22] analyzed the return and volatility connectedness of the stock prices of US clean energy companies, crude oil prices, and important financial variables in time and frequency domains.

We use a rolling analysis to examine the spillover of return and volatility in North America and Europe separately in time and frequency domains. Zhang and Wang [23] also analyzed the return and volatility spillover between the Chinese and global oil markets and employed a moving-window analysis to better understand and capture the dynamics of return and volatility spillover in time domain.

Finally, we investigate some prior studies similar to ours. Similar to our study, Oberndorfer et al. [24] focused on investigating the volatility spillover across energy markets and the pricing of European energy stocks by using the GARCH model, and they found that oil price is the main index for energy price developments in the European stock market. Kenourgios et al. [25] investigated the contagion effects of the global financial crisis (2007–2009) across assets in different regions. Kenourgios et al. [26] also investigated the contagion effects of the global financial crisis (2007–2009) in six developed and emerging regions by applying the FIAPARCH model. Baur [27] also studied different channels of financial contagions across 25 major countries and found that the crisis significantly increased the co-movement of returns. Singh et al. [28] examined price and volatility spillovers in the stock markets of North America, Asia, and Europe and found that a greater regional influence exists among the Asian and European stock markets. Balli et al. [29] analyzed the return and volatility spillovers and their determinants in emerging Asian and Middle Eastern countries. They found that developed financial markets have significant spillover effects on emerging financial markets and shocks originated in the US play a dominant role.

3. Empirical Techniques

3.1. Diebold–Yilmaz Method

Our study employs the method proposed by Diebold and Yilmaz [1] for measuring spillover in a generalized VAR framework. This approach was designed to measure the connectedness concept built on the basis of the generalized forecast error variance decomposition (GFEVD) of a VAR approximating model. First, we conceived an N-variable VAR (p) model, as (1) below.

$$y_t = \sum_{i=1}^{p} \Phi_i y_{t-i} + \varepsilon_t \tag{1}$$

where y_t is the N × 1 vector of the observed variables at time t, and Φ is the N × N coefficient matrix. The error vector ε_t is independent and identically distributed, and white noise $(0, \Sigma)$ with covariance matrix Σ is possibly non-diagonal.

In this model, the VAR process can also transform into the vector moving average (VMA) (∞), as represented in (2). It is effective to use with the (N × N) matrix lag-polynomial $|I_n - \Phi_1 z - \cdots - \Phi_p z^p| = 0$ with the I_n identity matrix. Assuming that the roots of $|\Phi(z)|$ lie outside the unit circle,

$$y_t = \psi(L)\varepsilon_t \tag{2}$$

where $\psi(L)$ is the (N × N) matrix of infinite lag polynomials that can be calculated from $\psi(L) = [\psi(L)]^{-1}$. However, as the order of the variables in the VAR system may influence the impulse response or variance decomposition results, to eliminate the influence from the ordering of the variables in the variance decomposition, Diebold and Yilmaz [1] applied the generalized VAR framework developed by Koop et al. [30] and Pesaran and Shin [31]. On the basis of this framework, the H-step-ahead GFEVD can be written in the form of (3):

$$(\theta_H)_{jk} = \frac{\sigma_{kk}^{-1} \sum_{h=0}^{H} ((\psi_h \Sigma)_{jk})^2}{\sum_{h=0}^{H} (\psi_h \Sigma \psi_h')_{jj}} \tag{3}$$

where ψ_h is an N × N coefficient matrix of polynomials at lag h, and $\sigma_{kk}^{-1} = (\Sigma)_{kk}$. $(\theta_H)_{jk}$ indicates the contribution of the kth variable of the model to the variance of the forecast error of element j at horizon h. To sum the elements in each row of the GFEVD to total 1, each entry is normalized by the row sum as

$$\tilde{\theta}_{jk}^H = \frac{\theta_{jk}^H}{\sum_{K=1}^{N} \theta_{jk}^H} \tag{4}$$

$\tilde{\theta}_{jk}^H$ measures the pairwise spillover from k to j at horizon H and also measures the spillover effect from market k to j. We can aggregate this to measure the total spillover of the system. The total spillover can be measured by the pairwise spillover. The connectedness can be seen as the share of variance in the forecasts contributed to by errors (Diebold and Yilmaz [1]).

$$S^H = 100 \times \frac{\sum_{j \neq k}^{N} \tilde{\theta}_{jk}^H}{\sum \tilde{\theta}^H} = 100 \times \left(1 - \frac{Tr\{\tilde{\theta}^H\}}{\sum \tilde{\theta}^H}\right) = 100 \times \left(1 - \frac{Tr\{\tilde{\theta}^H\}}{N}\right) \tag{5}$$

where $Tr\{.\}$ is the trace operator. The total spillover calculates the total spillover across all markets in the form of (5). There are two measures by Diebold and Yilmaz [1] that show the relative importance of each variable in the system:

Directional Spillover (From): $S_{k \leftarrow}^H = 100 \times \frac{\sum_{j=1, j \neq i}^{N} \tilde{\theta}_{kj}^H}{N}$; the directional spillover (from) is the spillover that market k receives from all other markets.

Directional Spillover (To): $S_{\leftarrow k}^H = 100 \times \frac{\sum_{j=1, j \neq i}^{N} \tilde{\theta}_{kj}^H}{N}$; the directional spillover (to) is the spillover that market k transmits to all other markets.

3.2. Barunik and Krehlik Method

Following Barunik and Krehlik [2], we describe the frequency dynamics (for the short-term, the medium-term, and the long-term) of spillover and the spectral formulation of variance decomposition.

Notably, we measure spillovers in frequency domain using Fourier transform. Moreover, the frequency response function is obtained as a Fourier transform of the coefficients ψ_h: $\psi(e^{-i\omega}) = \Sigma_h e^{-i\omega h}\psi_h$, where $i = \sqrt{-1}$. The generalized causation spectrum over frequencies $\omega \in (-\pi, \pi)$ is defined in the form of (6)

$$(f(\omega))_{jk} = \frac{\sigma_{kk}^{-1}\left|\left(\psi(e^{-i\omega})\Sigma\right)_{jk}\right|^2}{\left(\psi(e^{-i\omega})\Sigma\psi'(e^{+i\omega})\right)_{jj}} \quad (6)$$

where $\psi(e^{-i\omega}) = \Sigma_h e^{-i\omega h}\psi_h$ is the Fourier transform of the impulse response ψ_h. It is vital to pay attention to $(f(\omega))_{jk}$, namely, the portion of the spectrum of the jth variable at a given frequency ω due to shocks in the kth variable. As domination holds the spectrum of the jth variable at a given frequency ω, we establish (6) for the quantity within the frequency causation. To obtain the generalized decomposition of the variance's decompositions to frequencies, we weight $(f(\omega))_{jk}$ by the frequency share of the variance of the jth variable. This weighting function can be defined as (7)

$$\Gamma_j(\omega) = \frac{\left(\psi(e^{-i\omega})\Sigma\psi'(e^{+i\omega})\right)_{jj}}{\frac{1}{2\pi}\int_{-\pi}^{\pi}\left(\psi(e^{-i\lambda})\Sigma\psi'(e^{+i\lambda})\right)_{jj}d\lambda} \quad (7)$$

where the power of the jth variable at a given frequency sums through the frequencies to a constant value of 2π. When the Fourier transform of the impulse is a complex number value, the generalized factor spectrum is the squared coefficient of the weighted complex numbers and, therefore, a real number. In sum, we set the frequency band $d = (a, b)$: $a, b \in (-\pi, \pi), a < b$.

The GFEVD under the frequency band d is

$$\theta_{jk}(d) = \frac{1}{2\pi}\int_a^b \Gamma_j(\omega)(f(\omega))_{jk}d\omega \quad (8)$$

However, GFEVD will still be normalized into (9). The scaled GFEVD on the frequency band $d = (a, b)$: $a, b \in (-\pi, \pi), a < b$ is shown below:

$$\tilde{\theta}_{jk}(d) = \frac{\theta_{jk}(d)}{\Sigma_k \theta_{jk}(\infty)} \quad (9)$$

where $\tilde{\theta}_{jk}(d)$ is defined as the pairwise spillover at a given frequency band d. At the same time, it is possible to define the total spillover at frequency band d.

The frequency total spillover (frequency connectedness) on frequency band d can be defined as

$$S^F(d) = 100 \times \left(\frac{\Sigma\tilde{\theta}(d)}{\Sigma\tilde{\theta}(\infty)} - \frac{Tr\{\tilde{\theta}(d)\}}{\Sigma\tilde{\theta}(\infty)}\right) \quad (10)$$

where $Tr\{.\}$ is the trace operator, and $\Sigma\tilde{\theta}(d)$ is the sum of all elements of the $\tilde{\theta}(d)$ matrix. The frequency total spillover decomposes the total spillover into the long-term, the medium-term, and the short-term, and these can sum into the total spillover S, as defined by Diebold and Yilmaz [1].

Equally, we can also define the two directional spillovers for frequency according to Diebold and Yilmaz [1].

Frequency Directional Spillover (From): $S^F_{k\leftarrow}(d) = 100 \times \dfrac{\sum_{j=1, j\neq i}^{N} \tilde{\theta}_{kj}(d)}{N}$; the frequency directional spillover (from) is the spillover that market k receives from all other markets at frequency band d.

Frequency Directional Spillover (To): $S^H_{\leftarrow k}(d) = 100 \times \dfrac{\sum_{j=1, j\neq i}^{N} \tilde{\theta}_{kj}(d)}{N}$; the frequency directional spillover (to) is the spillover that market k transmits to all other markets at frequency band d.

4. Data

We use daily data for the natural gas, crude oil, and electricity utility sector indices of Europe and North America from 4 August 2009 to 16 August 2019, without uncommon business days. Because the data and CAC Utilities Index (USD) extends from 4 August 2009 to 16 August 2019, in order to consolidate the time for all data, we chose this time period. To avoid the influence of the exchange rate on our results, we consolidated the currency units of the variables into a dollar currency unit. Specifically, the variables we use are shown in Table 1.

Table 1. Variables in the model.

Variable	Data	Data Source
North America		
Natural Gas	Henry Hub Natural Gas Futures (USD/Million Btu)	Bloomberg
Crude Oil	Crude Oil WTI Futures (USD/Barrel)	Bloomberg
US	S&P 500 Utilities Index (USD)	Dow Jones
Canada	S&P/TSX Capped Utilities Sector Index (USD)	Bloomberg
Europe		
Natural Gas	ICE UK Natural Gas Futures (NBP) (USD/Million Btu)	Bloomberg
Crude Oil	ICE Brent Futures (USD/Barrel)	Bloomberg
Germany	DAX subsector All Electricity (USD)	DataStream
UK	FTSE 350 Electricity Index (USD)	DataStream
France	CAC Utilities Index (USD)	DataStream
Italy	FTSE ITALIA ALL-SHARE UTILITIES (USD)	DataStream

For the North American market, we use the daily prices of the Henry Hub Natural Gas Futures from Bloomberg for the natural gas market. For the crude oil market, we employ the Crude Oil WTI Futures from Bloomberg. With exports grown in Europe, South America, and Asia, the New York Mercantile Exchange (NYMEX) Henry Hub Natural Gas futures have become a global price benchmark for natural gas trading. WTI, a medium crude oil and futures contract launched by the NYMEX in 1983, has long been a benchmark for international crude oil prices, thus making an important contribution to the development of the global crude oil market. For the stock market, we use the electricity utilities in the US and Canada. For the US, we use the Standard and Poors 500 Utilities (S&P 500) which is composed of electricity and energy companies included in the S&P 500. This is classified as members of the Global Industry Classification Standard (GICS) utilities sector, such as the American Electric Power Company (AEP), Duke Energy Company (DU), Consolidated Edison Company (ED), and 28 other companies in total. The S&P/TSX Capped Utilities Sector Index (TSX) is a market-value-weighted index obtained from 16 electricity and energy companies, such as Emera Inc. (EMA) and Fortis Inc. (FTS).

For Europe, we use the daily price of the Intercontinental Exchange (ICE) UK Natural Gas Futures for the natural gas market and the ICE Brent Futures for the crude oil market. The ICE UK Natural Gas Futures contract is used for physical delivery by transfer of natural gas rights at the National Balancing Point (NBP) virtual trading point operated by the National Grid, a UK transmission system operator.

This is the second most common liquid gas trading point in Europe. Brent Oil is the primary trading category for sweet light crude and serves as a benchmark price for oil purchases in the world. For stock markets, in Germany, we use the Dax subsector All Electricity (DAX), calculated by Deutsche Börse. DAX is a market-value-weighted index that includes companies with an average daily trading volume of at least €1 million to qualify. In the UK, we use the market-value-weighted FTSE 350 Electricity Index (FTSE 350), which includes three companies: DRAXGROUP, SSE (Scottish and Southern Electricity), and Contour global, all of which are large electricity enterprises in the UK. In France, we use the CAC Utilities Index (CAC), a market-value-weighted index that comprises 10 electricity and energy companies, including EDF and ENGIE. In Italy, we use the FTSE ITALIA ALL-SHARE UTILITIES Index (FTSE Italia), which includes 14 electricity and energy companies, such as ENEL (an Italian multinational energy company working in the field of power generation and distribution).

Figure 1a shows images of the prices of natural gas, crude oil, and the electricity utility sector indices in North America. Figure 1b illustrates the natural gas, crude oil, and electricity utility sector indices in Europe. As we can see, in the North America market, the price of the natural gas, crude oil, and electricity utility sector index in Canada follow similar trends and fluctuate violently. These trends dropped dramatically from 2014 to 2016 due to the international oil price crisis. In general, the electricity utility sector index in the US increased continuously from 2009 to 2019, after the global financial crisis of 2007–2009. Relatively, in the European market, Figure 1b shows that the prices of natural gas, crude oil, and the electricity utility sector index follow a semblable non-stationary trend. Moreover, due to international oil price crisis of 2014–2016, energy prices dropped to varying degrees during this time period.

In our analysis, we calculated the closing price of the North American market and the European logarithmic difference as the daily return data, as shown in Figure 2a,b. Furthermore, we use the Ljung-Box, which has a lag of 20 to test the time variations of the return series and confirm that the return of all variables is not a white noise series with 10% significance. We use the ARMA (Autoregressive Moving Average)–GARCH model to calculate the volatilities of four assets in North America and six assets in Europe, and the plots are shown in Figure 3a,b. Additionally, the lag of the GARCH model is determined on the basis of the Akaike Information Criterion (AIC).

The descriptive basic statistics of the return and volatility series are shown in Table 2, Table 3. In North America, we find that the mean returns of natural gas and crude oil have negative values. However, the others have positive mean returns. Furthermore, natural gas has the largest maximum daily return and the largest minimum daily return. Specifically, we find that natural gas is the most volatile, followed by crude oil, TSX (Canada), and the S&P 500 (US). Moreover, based on skewness, we find that, except for natural gas, the returns are left-skewed, whereas the return of natural gas is right-skewed. Meanwhile, according to kurtosis, the volatility of the four assets are right-skewed. Regarding kurtosis, the returns and volatilities of the four assets are leptokurtic, which means that the four variables will show more peaked and fat tails. Finally, as the most commonly used unit root testing method, the augmented Dickey–Fuller (ADF), proposed in 1981, tests the null hypothesis that a variable has a unit root, which means that the variable is nonstationary. From the ADF results, the null hypothesis is rejected at 1% significance level for all variables.

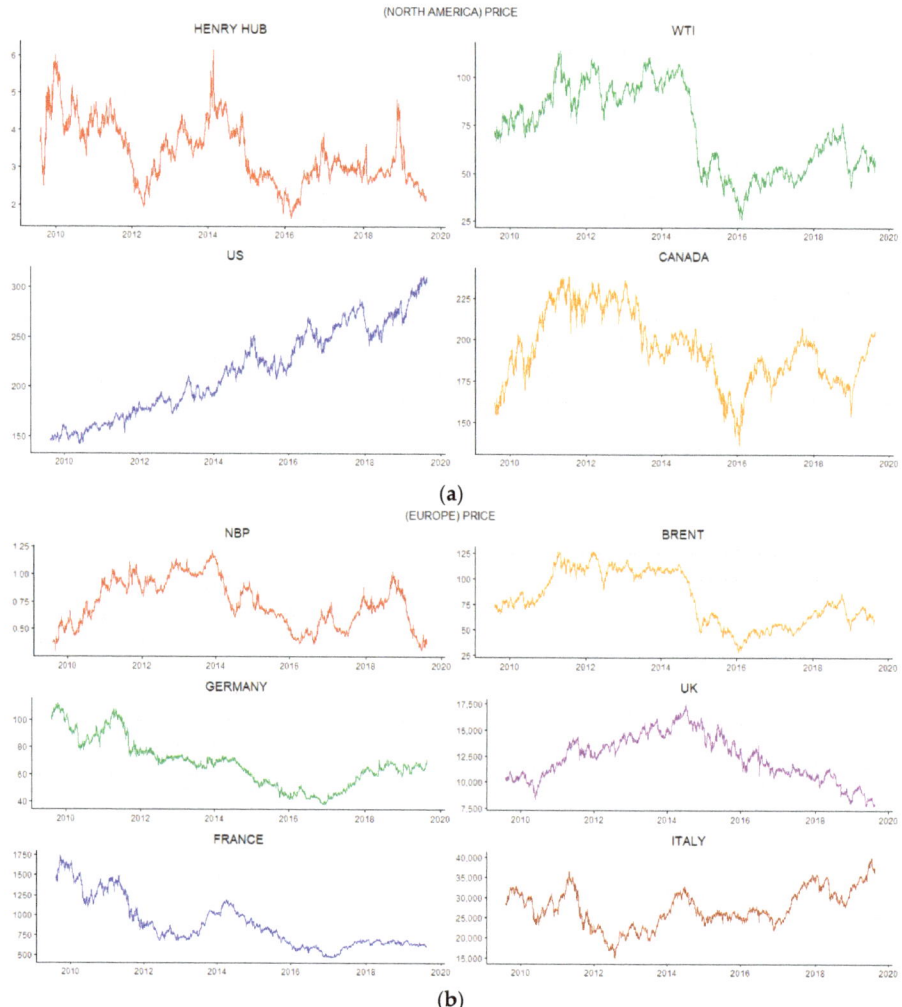

Figure 1. (**a**) Time-variations of the price series in North America. Notes: Henry Hub indicates natural gas; WTI indicates crude oil; US indicates the S&P 500 Utilities Index; CANADA indicates the S&P/TSX Capped Utilities Sector Index. (**b**) Time-variations of the price series in Europe. Notes: NBP indicates natural gas; BRENT indicates crude oil; GERMANY indicates the DAX subsector for all electricity; UK indicates the FTSE 350 Electricity Index; FRANCE indicates the CAC Utilities Index; ITALY indicates FTSE Italia All-Share Utilities Index.

Meanwhile, in the European market, in contrast to the North America market, we see that the mean returns of crude oil, DAX (Germany), FTSE 350 (UK), and CAC (France) have negative values. However, the others have positive mean returns. Natural gas has the largest maximum daily return, and FTSE 350 (the UK) has the largest minimum daily return. However, we find that natural gas is the most volatile, followed by crude oil, DAX (Germany), CAC (France), FTSE Italia (Italy), FTSE 350 (UK), and DAX (Germany). On the basis of skewness, the returns of natural gas and crude oil are right-skewed, but the others are left-skewed. In addition, the volatilities of the six assets are right-skewed. Regarding kurtosis, except for the return of CAC (France), which is platykurtic, the returns of the others are leptokurtic. Moreover, the volatilities of the six assets are leptokurtic. Eventually, based on the ADF results, we reject the null hypothesis of a unit root at 1% significance level for all series.

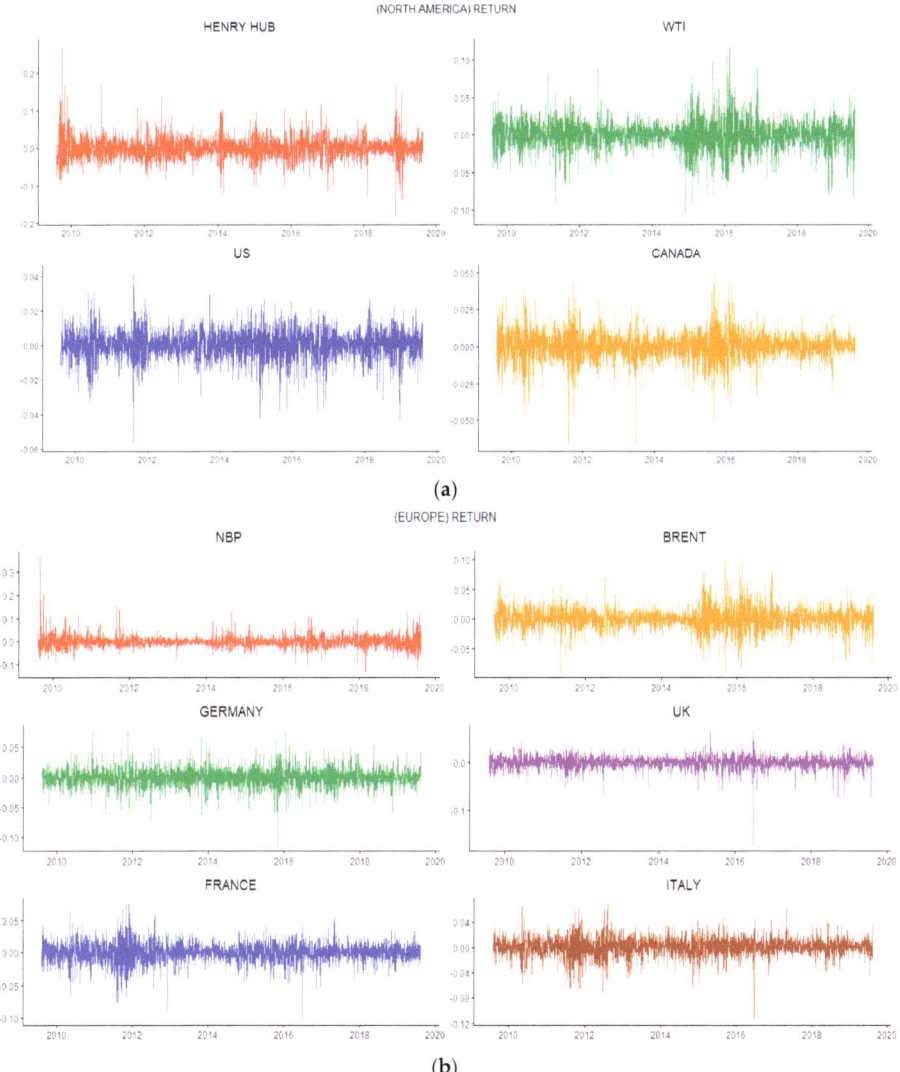

Figure 2. (**a**) Time-variations of the return series in North America. Notes: Henry Hub indicates natural gas; WTI indicates crude oil; US indicates the S&P 500 Utilities Index; CANADA indicates the S&P/TSX Capped Utilities Sector Index. (**b**) Time-variations of return series in Europe. Notes: NBP indicates natural gas; BRENT indicates crude oil; GERMANY indicates the DAX subsector for all electricity; UK indicates the FTSE 350 Electricity Index; FRANCE indicates the CAC Utilities Index; ITALY indicates FTSE Italia All-Share Utilities Index.

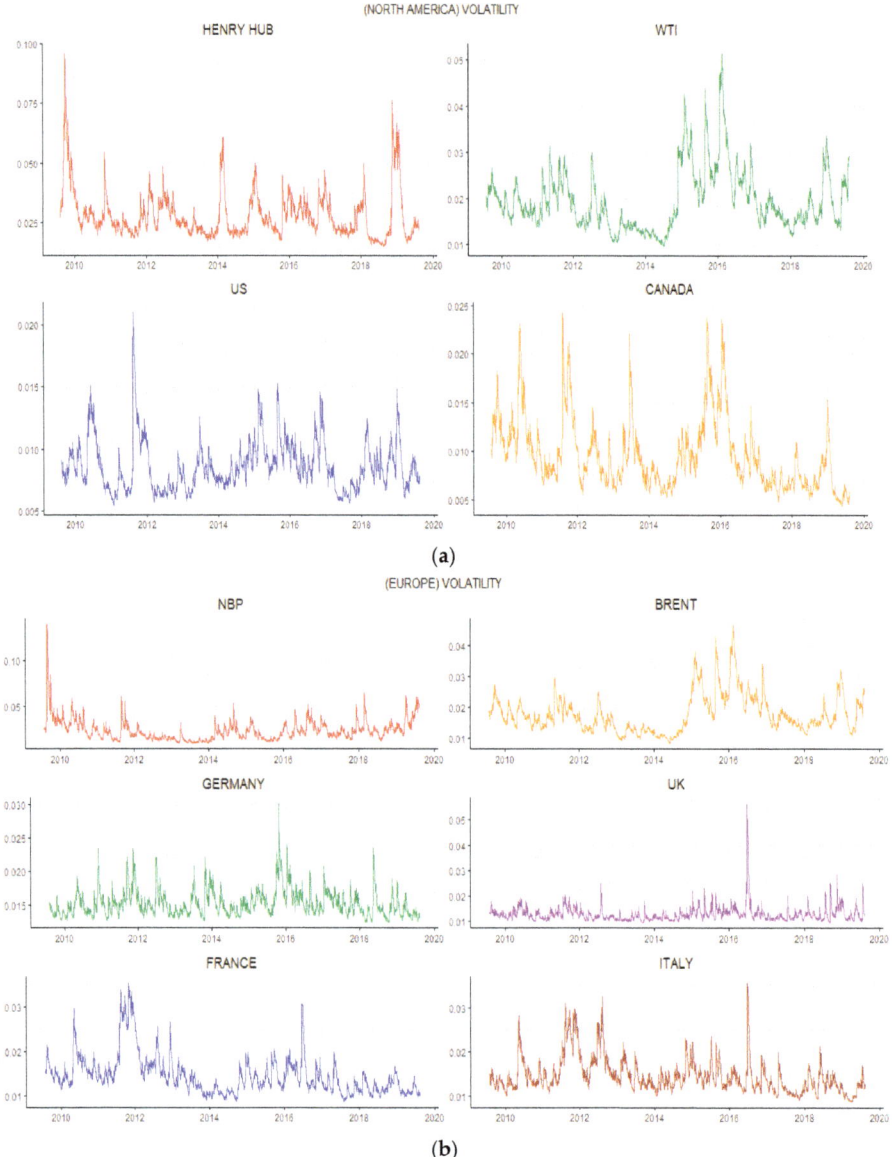

Figure 3. (a) Time-variations of the volatility series in North America. Notes: Henry Hub indicates natural gas; WTI indicates crude oil; US indicates the S&P 500 Utilities Index; CANADA indicates the S&P/TSX Capped Utilities Sector Index. (b) Time-variations of volatility series in Europe. Notes: NBP indicates natural gas; BRENT indicates crude oil; GERMANY indicates the DAX subsector for all electricity; UK indicates the FTSE 350 Electricity Index; FRANCE indicates the CAC Utilities Index; ITALY indicates FTSE Italia All-Share Utilities Index.

Table 2. Descriptive statistics of the return and volatility series in North America.

	Descriptive Statistics for Return			
	HENRY.HUB	WTI	US	CANADA
Mean	−0.00045	−0.00012	0.00029	0.00007
Median	−0.00080	0.00060	0.00079	0.00020
Maximum	0.268	0.116	0.041	0.049
Minimum	−0.181	−0.108	−0.056	−0.065
Std. Deviation	0.030	0.021	0.009	0.010
Skewness	0.567	−0.051	−0.465	−0.209
Kurtosis	6.008	2.758	2.292	3.193
ADF	−35.791 ***	−35.310 ***	−36.386 ***	−34.565 ***
	Descriptive Statistics for Volatility			
	HENRY.HUB	WTI	US	CANADA
Mean	0.028	0.020	0.009	0.010
Median	0.025	0.018	0.008	0.009
Maximum	0.096	0.051	0.021	0.024
Minimum	0.015	0.010	0.005	0.004
Std. Deviation	0.010	0.007	0.002	0.004
Skewness	2.055	1.346	1.416	1.346
Kurtosis	5.789	2.297	3.503	1.791
ADF	−5.446 ***	−3.954 ***	−5.658 ***	−4.710 ***

Note: ADF: Augmented Dickey and Fuller Unit Root Test (1979); *** denotes a rejection of the null hypothesis at 1% significance level.

Table 3. Descriptive statistics for return and volatility in Europe.

	Descriptive Statistics for Return					
	NBP	BRENT	GERMANY	UK	FRANCE	ITALY
Mean	0.00002	−0.00002	−0.00023	−0.00012	−0.00034	0.00014
Median	−0.00050	0.00040	−0.00030	0.00042	0.00000	0.00080
Maximum	0.363	0.104	0.078	0.066	0.074	0.068
Minimum	−0.131	−0.090	−0.114	−0.172	−0.101	−0.114
Std. Deviation	0.026	0.019	0.016	0.013	0.015	0.015
Skewness	1.687	0.005	−0.067	−1.382	−0.319	−0.454
Kurtosis	20.731	2.694	3.268	13.807	−0.319	3.205
ADF	−36.424 ***	−35.247 ***	−38.437 ***	−35.606 ***	−35.442 ***	−36.552 ***
	Descriptive Statistics for Volatility					
	NBP	BRENT	GERMANY	UK	FRANCE	ITALY
Mean	0.024	0.018	0.015	0.013	0.015	0.015
Median	0.022	0.017	0.015	0.012	0.014	0.014
Maximum	0.139	0.047	0.030	0.056	0.035	0.036
Minimum	0.010	0.008	0.012	0.010	0.009	0.009
Std. Deviation	0.012	0.006	0.002	0.003	0.004	0.004
Skewness	2.574	1.292	1.744	5.502	1.843	1.671
Kurtosis	13.910	2.021	4.675	56.087	4.066	3.403
ADF	−2.942 ***	−3.811 ***	−8.625 ***	−10.936 ***	−4.692 ***	−6.305 ***

Note: ADF: Augmented Dickey and Fuller Unit Root Test (1979); *** denotes a rejection of the null hypothesis at 1% significance level.

5. Empirical Results

5.1. Spillover Results

We followed Diebold and Yilmaz [1] (hence, DY12) to obtain the spillover effects in the time domain. We first estimated the four variables in our VAR model in the North American market and the six variables in our VAR model in the European market. Then, we used the generalized variance decomposition for a forecast error to set up the spillover table to measure the direction and intensity of spillover across the selected markets.

Next, in frequency domain, following Barunik and Krehlik [2] (hence, BK18), we used Fourier transform to decompose the spillover table in the DY12 model into three different frequency bands, stated in Toyoshima and Hamori [20] as the short-term, "Fre S," roughly corresponding to 1 to 5 days; the medium-term, "Fre M," roughly corresponding to 6 to 21 days; and the long-term, "Fre L," roughly corresponding to 22 days to infinity.

In this study, the lag length of the VAR model is based on the AIC. According to our BK18 model, if the forecasting horizon is (H) < 100, the method is invalid. Consequently, we used a 100-day ahead forecasting horizon (H) for variance decomposition.

In Table 4, Table 5, we show the return spillover results from DY12 and BK18, which include the two markets: North America and Europe. Specifically, Tables 4 and 5 includes four sub-tables. At the top are the results of the DY12, followed by the short-term, the medium-term, and the long-term results from the BK18. In each sub-table, the values in the ith row and the jth column equate to the strength of the spillover effect from the jth market to the ith market. For example, in the DY12 return spillover results in the North American market, the strength of the spillover effect in the third column (US) and the second row (WTI) is 2.264. The values in the last row called "TO" indicate the mean value of the spillover effect on the other markets, whereas the values in the last column called "FROM" indicate the mean value of the spillover effect from the other markets. The total spillover is the summary of all "TO" or "FROM" (for example, 17.165, as shown in the lower right corner).

Table 4. The return spillover table of DY, 2012 and BK, 2016 (North America).

	North America (Return)				
	Spillover Results (DY12)				
	HENRY.HUB	WTI	US	CANADA	FROM
HENRY.HUB	98.033	1.425	0.147	0.394	0.492
WTI	1.177	82.863	2.264	13.697	4.284
US	0.163	2.242	79.294	18.301	5.177
CANADA	0.318	11.8	16.73	71.151	7.212
TO	0.415	3.867	4.785	8.098	17.165
	Spillover Results (BK18)				
	Frequency S 1–5 Days				
	HENRY.HUB	WTI	US	CANADA	FROM
HENRY.HUB	82.275	1.237	0.117	0.326	0.42
WTI	0.968	68.456	1.825	10.802	3.399
US	0.157	1.94	64.922	14.496	4.148
CANADA	0.227	9.137	12.591	55.607	5.489
TO	0.338	3.078	3.633	6.406	13.455
	Frequency M 6–21 Days				
	HENRY.HUB	WTI	US	CANADA	FROM
HENRY.HUB	11.643	0.14	0.022	0.05	0.053
WTI	0.154	10.627	0.322	2.127	0.651
US	0.005	0.223	10.591	2.796	0.756
CANADA	0.067	1.956	3.033	11.414	1.264
TO	0.056	0.58	0.844	1.243	2.724

Table 4. *Cont.*

	Frequency L 22–Infinite Days				
	HENRY.HUB	WTI	US	CANADA	FROM
HENRY.HUB	4.116	0.049	0.008	0.018	0.019
WTI	0.055	3.78	0.117	0.768	0.235
US	0.002	0.079	3.781	1.009	0.272
CANADA	**0.025**	**0.708**	1.106	4.13	0.46
TO	0.02	0.209	0.308	0.449	**0.986**

Note: Freq S: the spillover at "Freq S" roughly corresponds to 1 to 5 days; Freq M: the spillover at "Freq M" roughly corresponds to 6 to 21 days; Freq L: the spillover at "Freq L" roughly corresponds to 22 to infinite days.

Table 5. Return spillover table of DY (2012) and BK (2018) (Europe).

Europe (Return)							
Spillover Results (DY12)							
	NBP	BRENT	GERMANY	UK	FRANCE	ITALY	FROM
NBP	95.472	1.562	0.229	**1.249**	0.899	0.588	0.755
BRENT	1.476	80.502	0.859	5.142	**6.353**	5.667	3.25
GERMANY	0.315	1.176	85.497	1.441	5.395	6.177	2.417
UK	**0.757**	**3.889**	1.016	59.907	18.122	16.309	6.682
FRANCE	0.487	3.882	3.076	14.526	48.209	29.82	8.632
ITALY	0.353	3.452	3.568	13.401	30.282	48.944	8.509
TO	0.565	2.327	1.458	5.96	10.175	9.76	**30.245**
Spillover Results (BK18)							
Frequency S 1–5 Days							
	NBP	BRENT	GERMANY	UK	FRANCE	ITALY	FROM
NBP	77.614	1.2	0.177	0.926	0.656	0.456	0.569
BRENT	1.364	66.891	0.795	4.062	5.013	4.614	2.641
GERMANY	0.31	0.797	71.894	1.123	4.185	4.976	1.898
UK	**0.624**	**2.975**	0.799	48.48	14.355	13.195	5.325
FRANCE	0.452	2.943	2.373	11.673	38.348	23.9	6.89
ITALY	0.332	2.792	2.84	11.254	24.717	40.276	6.989
TO	0.514	1.785	1.164	4.84	8.154	7.857	**24.313**
Frequency M 6–21 Days							
	NBP	BRENT	GERMANY	UK	FRANCE	ITALY	FROM
NBP	13.161	0.266	0.038	0.236	0.178	0.097	0.136
BRENT	0.085	10.047	0.048	0.794	0.985	0.775	0.448
GERMANY	0.005	0.277	10.052	0.233	0.889	0.884	0.381
UK	**0.098**	0.67	0.16	8.414	2.768	2.292	0.998
FRANCE	0.027	**0.688**	0.517	2.099	7.252	4.357	1.281
ITALY	0.016	0.486	0.536	1.587	4.102	6.393	1.121
TO	0.038	0.398	0.216	0.825	1.487	1.401	**4.365**
Frequency L 22–Infinite Days							
	NBP	BRENT	GERMANY	UK	FRANCE	ITALY	FROM
NBP	4.697	0.097	0.014	0.087	0.065	0.035	0.05
BRENT	0.028	3.564	0.016	0.286	0.355	0.278	0.16
GERMANY	0.001	0.101	3.551	0.084	0.322	0.317	0.137
UK	**0.035**	0.243	0.058	3.013	0.999	0.822	0.359
FRANCE	0.008	**0.25**	0.187	0.753	2.61	1.564	0.46
ITALY	0.005	0.174	0.192	0.56	1.463	2.275	0.399
TO	0.013	0.144	0.078	0.295	0.534	0.503	**1.567**

Note: Freq S: the spillover at "Freq S" roughly corresponds to 1 to 5 days; Freq M: the spillover at "Freq M" roughly corresponds to 6 to 21 days; Freq L: the spillover at "Freq L" roughly corresponds to 22 to infinite days.

Next, we identify some specifics when we compare the difference between the return spillover effect in North America and in Europe using the pure time-domain approach of DY12. First, for the total return spillover effect, we can see that the total return spillover effect in Europe (30.245%) is stronger than that in North America (17.165%). This demonstrates that the return connectedness of natural gas, crude oil, and the electricity utility sector index in Europe is stronger than that in North America. Whether in North America or Europe, the total return spillover of crude oil contributes more than natural gas to the electricity utility stock market in each country. Further, in North America, Canada receives a greater effect from the two energy futures (0.318% from natural gas and 11.8% from crude oil) compared with the US in the time domain. In contrast, Canada exerts a greater effect on the natural gas (0.394%) and crude oil (13.697%) sectors in the time domain. In Europe, the UK receives the greatest effect from the two energy futures (0.757% from natural gas and 3.889% from crude oil) compared with the other three countries in the time domain. The UK transmits the greatest effect on natural gas (1.249%), but France transmits the greatest effect on crude oil (6.353%) in the time domain, which contrasts with the situation in North America. Finally, we find that there are some differences in the return spillover effect among the two commodity markets and the stock market in North America and Europe, respectively. In North America, we find that the US receives a greater return spillover effect on natural gas (0.163%) and transmits less of an effect (0.147%). In contrast to the US, Canada transmits more of a return spillover effect on natural gas (0.394%) and receives a smaller effect (0.318%). In contrast to the natural gas market, the US transmits more of a return spillover effect on crude oil (2.264%) and receives a smaller effect (2.242%). In the same vein as the US, Canada also transmits more of a return spillover effect on crude oil (13.697%) and receives a smaller effect (11.8%).

In Europe, except for Germany, the other three stock markets transmit more of a return spillover effect on natural gas and receive a smaller effect. In the same vein as the situation for natural gas, except for Germany, the other three stock markets transmit more of a return spillover effect on crude oil and receive a smaller effect. In terms of frequency, Tables 4 and 5 reveal that looking at either North America or Europe, the total return spillover in the short-term (Frequency S 1–5 Days) is the highest, followed by the medium-term (Frequency M 6–21 Days) and the long-term (Frequency L 22–Infinity Days). These results suggest that the return shocks from any market transmitted to another market will not exceed one week.

Table 6, Table 7 show the volatility spillover results of DY12 and BK18 with the same construction. Table 6 shows the volatility spillover effect in North America, and Table 7 shows the effect in Europe. In time domain of DY12, we find the following. The total volatility spillover effect in Europe (27.929%) is stronger than that in in North America (20.216%). This situation is the same for the total return spillover; whether in North America or Europe, the total volatility spillover of crude oil transmits more than natural gas to the electricity utility stock market in each country. In North America, between the two stock markets, Canada receives the greatest spillover effect from natural gas (0.664%), and the US receives the greatest effect from crude oil (6.771%) in the time domain. Among the two commodity markets, WTI is the most influential market on the stock market, transmitting the largest volatility spillover to the US (6.771%) and Canada (6.302%) and is also the market that receives the largest volatility spillover from the US (4.373%) and Canada (19.036%) compared with the natural gas market.

In the European market, among the four stock markets, the UK receives the largest effect from natural gas (0.785%), and Germany receives the largest effect from crude oil (4.56%) in the time domain. However, the UK transmits the largest effect on crude oil (8.843%), and Germany transmits the largest effect on natural gas (2.374%) compared with other countries in the time domain. In addition, in terms of frequency and in contrast to the return spillover, Tables 6 and 7 reveal that in both North America and Europe, the total volatility spillover for the long-term (Frequency L 22–Infinity Days) is the highest, followed by the medium-term (Frequency M 6–21 Days) and the short-term (Frequency S 1–5 Days). These results imply that volatility shocks have a long-lasting effect.

Table 6. Volatility spillover table of DY12 and BK18 (North America).

	North America (Volatility)				
	Spillover Results (DY12)				
	HENRY.HUB	WTI	US	CANADA	FROM
HENRY.HUB	96.379	0.471	1.073	**2.077**	0.905
WTI	2.715	73.876	4.373	**19.036**	6.531
US	0.577	**6.771**	67.206	25.446	8.199
CANADA	**0.664**	6.302	11.36	81.674	4.581
TO	0.989	3.386	4.201	11.64	**20.216**
	Model Spillover Results (BK18)				
	Frequency S 1–5 Days				
	HENRY.HUB	WTI	US	CANADA	FROM
HENRY.HUB	2.19	0.001	0	0.03	0.008
WTI	0.011	0.686	0.006	0.001	0.004
US	0.01	0.012	1.871	0.144	0.042
CANADA	**0.015**	**0.029**	0.175	1.07	0.055
TO	0.009	0.01	0.045	0.044	**0.108**
	Frequency M 6–21 Days				
	HENRY.HUB	WTI	US	CANADA	FROM
HENRY.HUB	6.192	0.002	0.004	0.084	0.023
WTI	0.032	1.951	0.018	0.006	0.014
US	0.028	0.037	5.257	0.415	0.12
CANADA	**0.042**	**0.083**	0.498	3.036	0.156
TO	0.025	0.03	0.13	0.126	**0.312**
	Frequency L 22–Infinite Days				
	HENRY.HUB	WTI	US	CANADA	FROM
HENRY.HUB	87.998	0.469	1.068	1.963	0.875
WTI	2.673	71.239	4.35	19.029	6.513
US	0.539	**6.722**	60.078	24.886	8.037
CANADA	**0.607**	6.19	10.686	77.568	4.371
TO	0.955	3.345	4.026	11.47	**19.796**

Note: Freq S: the spillover at "Freq S" roughly corresponds to 1 to 5 days; Freq M: the spillover at "Freq M" roughly corresponds to 6 to 21 days; Freq L: the spillover at "Freq L" roughly corresponds to 22 to infinite days.

5.2. Dynamic (Moving-Window) Analysis

We have useful results on the total spillover effects from our full sample. However, these results are not helpful in analyzing how connectedness changes over time. If we only focus on the static results, the VAR estimated over the whole sample may smooth out the results when there is time variation in the relationship between the variables (Lovcha [21]). In order to better understand the dynamics of spillover effects, we employ a moving-window to analyze the spillover results of DY12 and BK18. For the moving-window method, we keep the forecast horizon at 100, which is used in the static analysis. For example, Toyoshima and Hamori [20] employ 100–day rolling samples. Jorion [32] sets a 20–day window for estimation. Blanchard et al. [33] used a five–year rolling standard deviation of output growth in the US. Similarly, we set the length of the window at 250 trading days, 370 trading days, and 500 trading days, and find that the plots of these trading days have almost the same trends. We put the results into Appendix A. For this reason, we chose 500 trading days for the length of the moving-window to keep the rolling sample large enough to ensure the stationarity of the series in each VAR estimation.

Table 7. Volatility spillover table of) DY12 and BK18 (Europe).

	\multicolumn{7}{c}{Europe (Volatility)}						
	\multicolumn{7}{c}{Spillover Results (DY12)}						
	NBP	BRENT	GERMANY	UK	FRANCE	ITALY	FROM
NBP	95.825	0.142	**2.374**	0.57	0.315	0.775	0.696
BRENT	7.628	72.585	0.516	**8.843**	3.489	6.939	4.569
GERMANY	0.372	**4.56**	85.188	0.815	4.103	4.962	2.469
UK	**0.785**	4.186	0.68	69.947	11.411	12.99	5.009
FRANCE	0.061	1.067	2.13	10.504	56.435	29.804	7.261
ITALY	0.252	0.382	2.676	9.658	34.583	52.448	7.925
TO	1.517	1.723	1.396	5.065	8.984	9.245	**27.929**
	\multicolumn{7}{c}{Spillover Results (BK18)}						
	\multicolumn{7}{c}{Frequency S 1–5 Days}						
	NBP	BRENT	GERMANY	UK	FRANCE	ITALY	FROM
NBP	4.619	0.003	0.001	0.001	0.001	0.002	0.001
BRENT	0.012	0.645	0	0.004	0.008	0.001	0.004
GERMANY	0.002	0.002	5.888	0.001	0.03	0.026	0.01
UK	**0.003**	0.013	0.003	7.448	0.889	0.892	0.3
FRANCE	0	0.02	0.004	0.101	0.9	0.34	0.078
ITALY	0.002	**0.037**	0.009	0.217	0.642	1.782	0.151
TO	0.003	0.012	0.003	0.054	0.262	0.21	**0.544**
	\multicolumn{7}{c}{Frequency M 6–21 Days}						
	NBP	BRENT	GERMANY	UK	FRANCE	ITALY	FROM
NBP	12.611	0.007	0.036	0.016	0.008	0.014	0.013
BRENT	0.042	1.839	0.002	0.042	0.033	0.027	0.024
GERMANY	**0.021**	0.017	15.534	0.025	0.072	0.063	0.033
UK	0.011	0.052	0.027	17.694	2.135	2.188	0.735
FRANCE	0.002	0.056	0.017	0.33	2.589	1.011	0.236
ITALY	0.005	**0.099**	0.043	0.677	1.849	4.933	0.446
TO	0.013	0.038	0.021	0.181	0.683	0.551	**1.487**
	\multicolumn{7}{c}{Frequency L 22–Infinite Days}						
	NBP	BRENT	GERMANY	UK	FRANCE	ITALY	FROM
NBP	78.595	0.133	2.337	0.553	0.306	0.758	0.681
BRENT	7.574	70.100	0.514	8.798	3.448	6.910	4.541
GERMANY	0.350	**4.541**	63.766	0.789	4.000	4.873	2.426
UK	**0.772**	4.121	0.650	44.805	8.387	9.910	3.973
FRANCE	0.059	0.991	2.109	10.073	52.945	28.453	6.947
ITALY	0.245	0.246	2.624	8.764	32.092	45.733	7.329
TO	1.500	1.672	1.372	4.829	8.039	8.484	**25.898**

Note: Freq S: the spillover at "Freq S" roughly corresponds to 1 to 5 days; Freq M: the spillover at "Freq M" roughly corresponds to 6 to 21 days; Freq L: the spillover at "Freq L" roughly corresponds to 22 to infinite days.

As displayed in Figure 4a,b, we find some characteristics from the dynamics of the total return spillover and the frequency decomposition of North America and Europe. In both North America and Europe, the total return spillover occurs in the short-term. Moreover, two dynamic return spillover figures have a similar trend. In Europe, the total spillover of the DY12 model for the return series varies between 20% and 45%, which is wider than that in North America (between 10% and 30%). From 2009 to mid–2013, whether in North America or Europe, the total return spillover retains high stability, which may be a consequence of the effect of the 2008 global financial crisis or the 2010 European sovereign debt crisis. The total return spillover in both North America and Europe began to increase steadily until around 2017, which could be influenced by the 2014 international crude oil crisis.

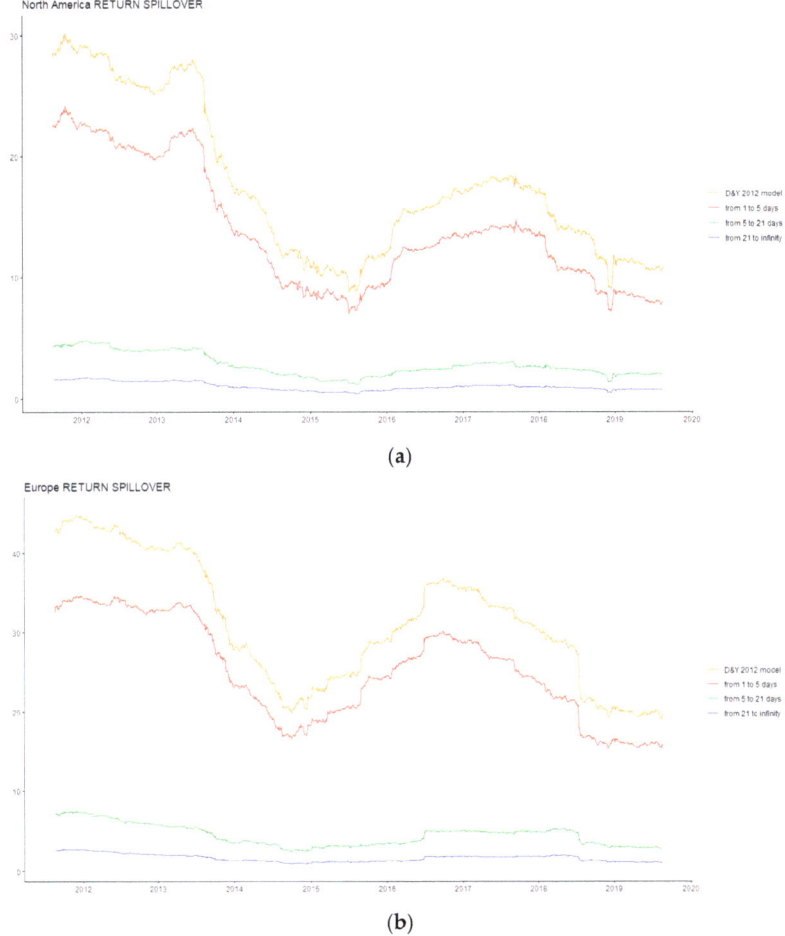

Figure 4. (a) The total return spillover of DY12 and BK18 in North America (Windows 500). (b) The total return spillover of DY12 and BK18 in Europe (Windows 500). Note: The yellow line indicates the total spillover index of the DY12 model; the red line indicates the total spillover index at "Freq S" of the BK18 model; the green line indicates the total spillover index at "Freq M" of the BK18 model; the blue line indicates the total spillover index at "Freq L" of the BK18 model. The vertical axis variable unit is in percentages.

As seen in Figure 5a,b, which shows the dynamics of total volatility spillover and frequency decomposition for North America and Europe, the total spillover fluctuates more than the total return spillover. Moreover, the total volatility spillover reacts more violently to extreme events than that of the returns. In contrast with the total return spillover, the total volatility spillover develops over the long-term. This means that the total volatility spillover is more sensitive to shocks and extreme events; unlike the total return spillover, whether in North America or Europe, the total volatility spillover of DY12 shifts between 0% and 60%. From 2009 to mid-2013, whether in North America or Europe, the total volatility spillover retained its stability, which may be a consequence of the 2008 global financial crisis and the 2009 European sovereign debt crisis. In North America, we identified some sudden fluctuations, such as a sharp increase in 2015 and 2016, an increasing trend from 2017 to 2018, and a sudden rise in 2019. These fluctuations may have been influenced by the 2014 international crude oil crisis, the 2016 Organization of the Petroleum Exporting Countries announcement to cut supplies at the end of 2017, the 2017 summer Hurricane Irma, and the 2019 trade war between China and the US. However, in Europe, we see almost the same fluctuations from mid-2013 to 2016; around 2015 to mid-2016, there are two sudden fluctuations that may be a result of some extreme events, such as the 2014 international crude oil crisis and the 2016 Brexit event. We also investigated the dynamics of the total return and volatility spillover in different windows and the pairwise directional return and volatility spillover in two regions. The results are given in Appendix A.

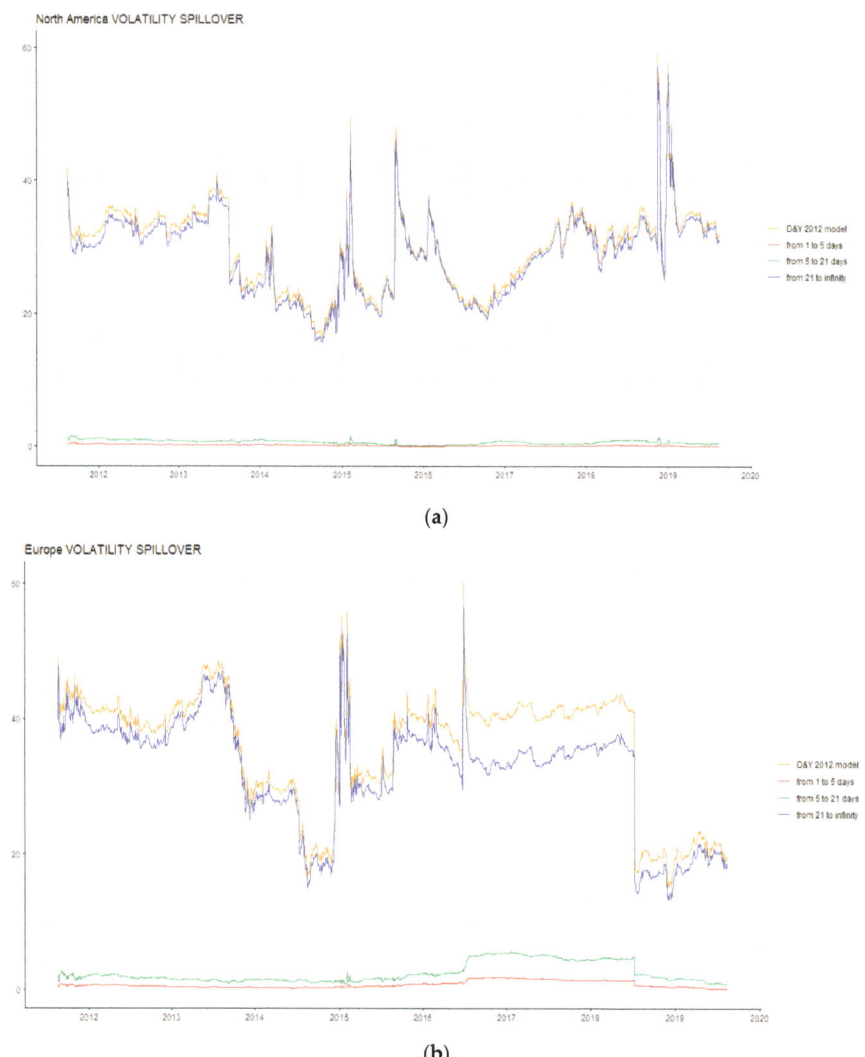

Figure 5. (a) The total volatility spillover of DY12 and BK18 in North America (Windows 500). (b) The total volatility spillover of DY12 and BK18 in Europe (Windows 500). Note: The yellow line indicates the total spillover index of the DY12 model; the red line indicates the total spillover index at "Freq S" of the BK18 model; the green line indicates the total spillover index at "Freq M" of the BK18 model; the blue line indicates the total spillover index at "Freq L" of the BK18 model. The vertical axis variable unit is in percentages.

6. Concluding Remarks

Our paper discusses the return and volatility spillover across the natural gas market, crude oil market, and stock market from 4 August 2009 to 16 August 2019 to assess the information transmission and risk transmission among the three markets by employing a new method for time–frequency developed by Diebold and Yilmaz [1] and Barunik and Krehlik [2]. It is crucial to investigate the spillover effects not only for investors to adjust their investment programs but also for government authorities to make proper economic decisions. The contributions of our paper to the literature are as follows.

First, in time domain, the total return and volatility spillover in Europe are stronger than in North America. Moreover, whether in North America or in Europe, the spillover table reveals that crude oil, rather than natural gas, has the greatest effect on electricity utility stock markets. In North America, Canada not only receives a larger return spillover effect from the two energy futures (0.318% from natural gas, 11.8% from crude oil) compared with the US but also transmits a greater effect on the two energy futures. Regarding volatility spillover, Canada still has the largest spillover effect on the other two energy futures and receives the largest volatility effect from natural gas. However, the US receives the largest volatility spillover effect from crude oil (6.771%). In Europe, the UK receives the greatest return spillover effect from the two energy futures (0.757% from natural gas and 3.889% from crude oil) compared with the other three countries. The UK transmits the largest effect on natural gas (1.249%), but France transmits the largest effect on crude oil (6.353%) in the time domain, which is different than the situation in North America. Regarding volatility spillover, the UK receives the largest effect from natural gas (0.785%), and Germany receives the largest from crude oil (4.56%). However, FTSE350 (the UK) transmits the largest effect on crude oil (8.843%), and Germany transmits the largest effect on natural gas (2.374%) compared to the other countries.

Second, in terms of frequency, our results show that the short-term has the largest effect on return spillover; however, the long-term has the largest effect on volatility spillover in both North America and Europe. These results imply that the return shocks from any market transmitted to another market will not exceed one week, whereas the results of volatility spillovers imply that volatility shocks have a long-lasting effect. This conclusion is consistent with the results of Barunik and Krehlik [2] and Tiwari et al. [34].

Third, in terms of return spillover transmission, all markets respond to return shocks immediately. The total return spillover effect in the short-term was the greatest. In this case, it is difficult to determine the impact of a particular market on another market. Unlike return spillover transmissions, the total volatility spillover effect in the long-term was the greatest. Policymakers have sufficient time to prevent the impact of extreme volatility shocks on other markets. In addition, based on a summary of the results above, the volatility of natural gas is less than that of oil, which suggests that compared with oil, natural gas investors may have a greater opportunity to make a profit.

Forth, some interesting results are displayed in the rolling analyses. For example, because of the subsequent effect of the 2008 global financial crisis and the 2010 European sovereign debt crisis, the return and volatility spillover in both North America and Europe maintained a high level. Due to the later 2014 international oil crisis, both North America and Europe fluctuated fiercely around 2015. Around mid-2016, the Brexit event made the volatility spillover in Europe increase suddenly.

Author Contributions: Conceptualization, T.N. and S.H.; investigation, W.Z. and X.H.; writing—original draft preparation, W.Z.; writing—review and editing, X.H., T.N., and S.H.; project administration, S.H.; funding acquisition, S.H. All authors have read and agreed to the published version of the manuscript.

Funding: This work was supported by JSPS KAKENHI Grant Number 17H00983.

Acknowledgments: We are grateful to Mehmet H. Bilgin and four anonymous referees for helpful comments and suggestions.

Conflicts of Interest: The authors declare no conflict of interest.

Appendix A

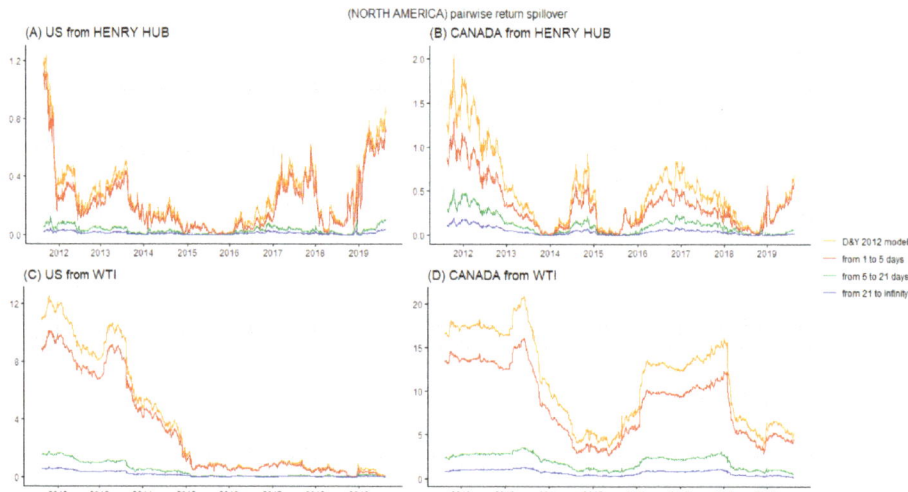

Figure A1. Pairwise directional return spillover of DY12 and BK18 in North America. Notes: US from HENRY HUB indicates the return spillover from natural gas to US; CANADA from HENRY HUB indicates the return spillover from natural gas to Canada; US from WTI indicates the return spillover from crude oil to US; CANADA from WTI indicates the return spillover from crude oil to Canada.

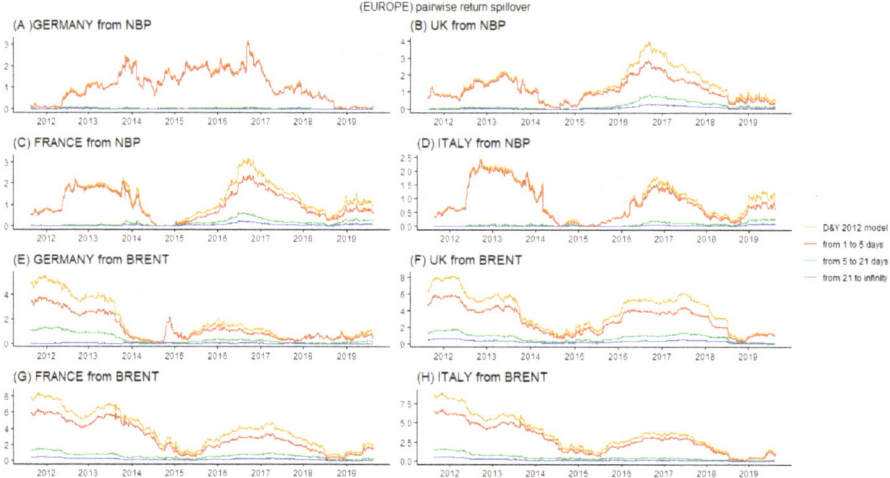

Figure A2. Pairwise directional return spillover of DY12 and BK18 in Europe. Notes: GERMANY from NBP indicates the return spillover from natural gas to Germany; UK from NBP indicates the return spillover from natural gas to the UK; FRANCE from NBP indicates the return spillover from natural gas to France; ITALY from NBP indicates the return spillover from natural gas to Italy; GERMANY from BRENT indicates the return spillover from crude oil to Germany; UK from BRENT indicates the return spillover from crude oil to the UK; FRANCE from BRENT indicates the return spillover from crude oil to France; ITALY from BRENT indicates the return spillover from crude oil to Italy.

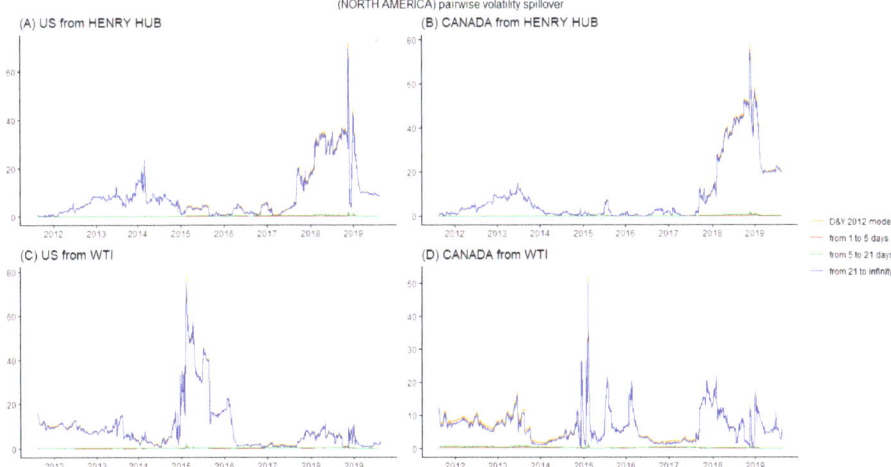

Figure A3. Pairwise directional volatility spillover of DY12 and BK18 in North America. Notes: US from HENRY HUB indicates the volatility spillover from natural gas to US; CANADA from HENRY HUB indicates the volatility spillover from natural gas to Canada; US from WTI indicates the volatility spillover from crude oil to US; CANADA from WTI indicates the volatility spillover from crude oil to Canada.

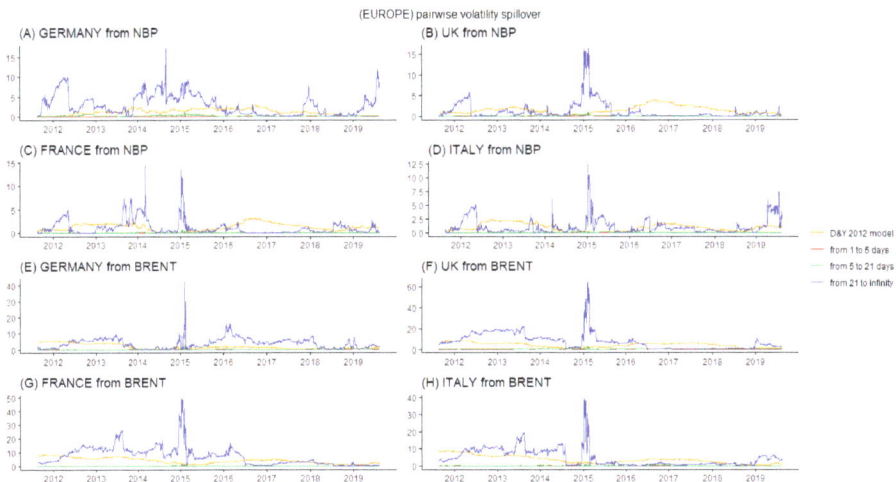

Figure A4. Pairwise directional volatility spillover of DY12 and BK18 in Europe. Notes: GERMANY from NBP indicates the volatility spillover from natural gas to Germany; UK from NBP indicates the volatility spillover from natural gas to the UK; FRANCE from NBP indicates the volatility spillover from natural gas to France; ITALY from NBP indicates the volatility spillover from natural gas to Italy; GERMANY from BRENT indicates the volatility spillover from crude oil to Germany; UK from BRENT indicates the volatility spillover from crude oil to the UK; FRANCE from BRENT indicates the volatility spillover from crude oil to France; ITALY from BRENT indicates the volatility spillover from crude oil to Italy.

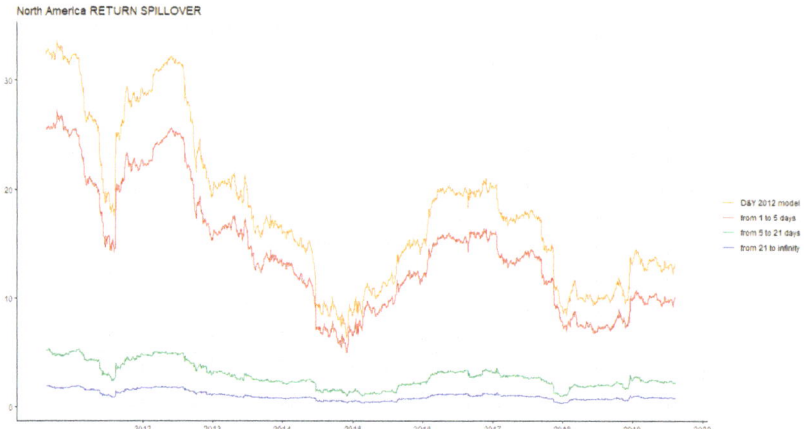

Figure A5. Total return spillover of DY12 and BK18 in North America (Windows 250).

Figure A6. Total return spillover of DY12 and BK18 in North America (Windows 370).

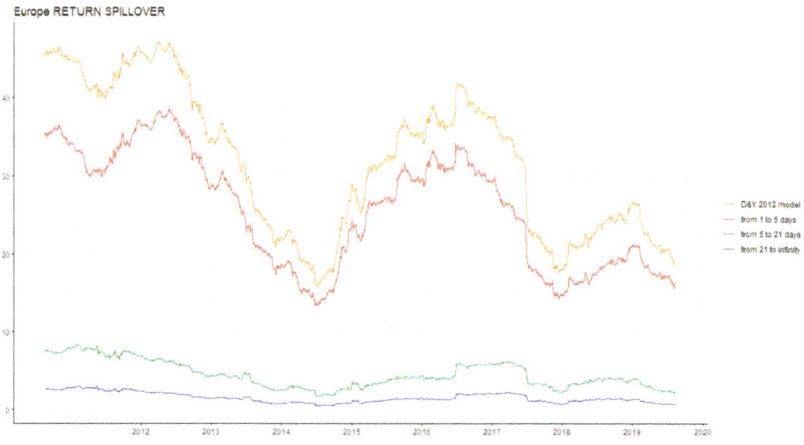

Figure A7. Total return spillover of DY12 and BK18 in Europe (Windows 250).

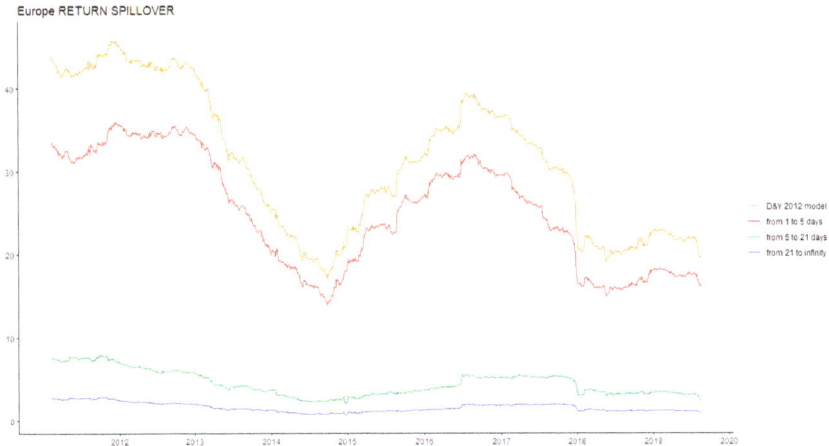

Figure A8. Total return spillover of DY12 and BK18 in Europe (Windows 370).

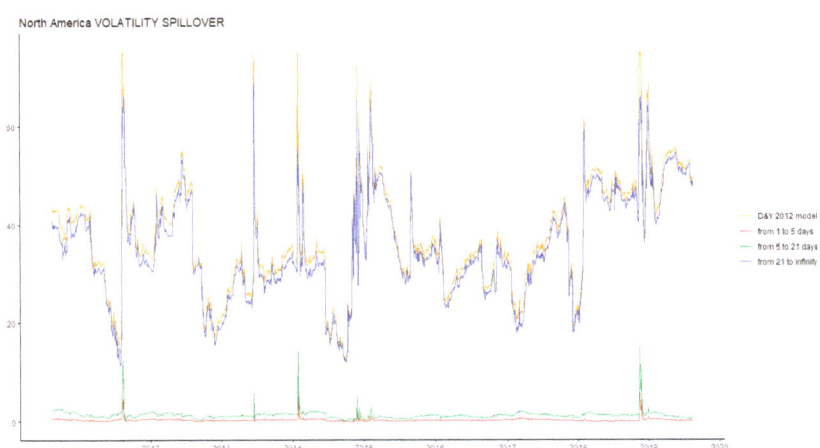

Figure A9. Total volatility spillover of DY12 and BK18 in North America (Windows 250).

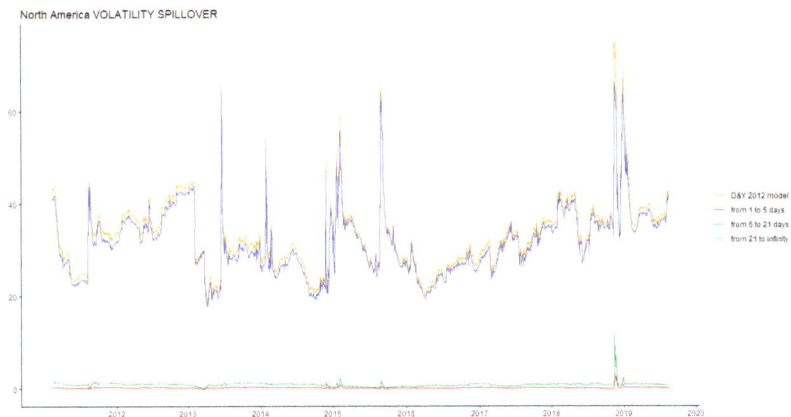

Figure A10. Total volatility spillover of DY12 and BK18 in North America (Windows 370).

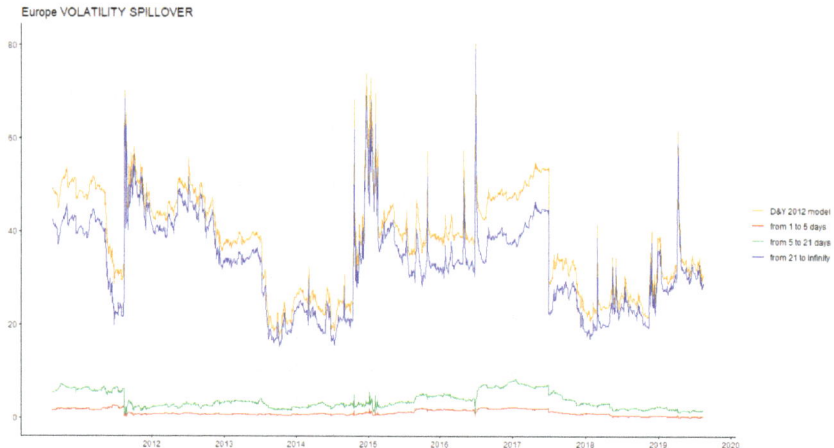

Figure A11. Total volatility spillover of DY12 and BK18 in Europe (Windows 250).

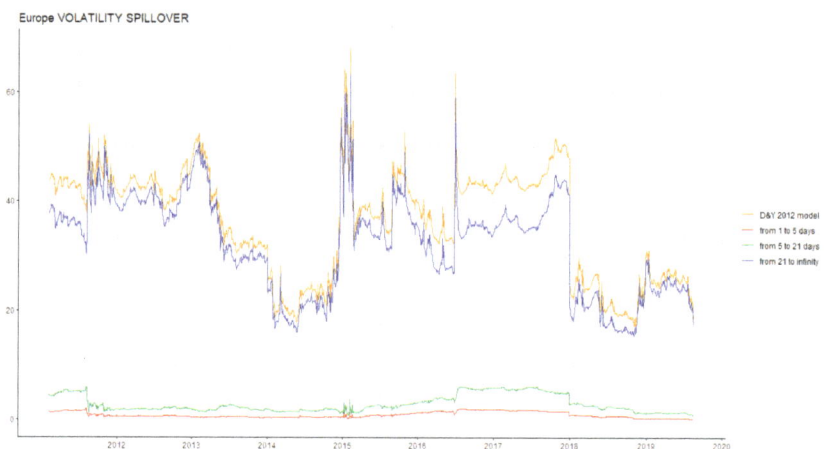

Figure A12. Total volatility spillover of DY12 and BK18 in Europe (Windows 370).

References

1. Diebold, F.; Yilmaz, K. Better to give than to receive: Predictive directional measurement of volatility spillover. *Int. J. Forecast.* **2012**, *28*, 57–66. [CrossRef]
2. Barunik, J.; Krehlik, T. Measuring the Frequency Dynamics of Financial Connectedness and Systemic Risk. *J. Financ. Econ.* **2018**, *16*, 271–296. [CrossRef]
3. Dudley, B. BP Statistical Review of World Energy. 2019. Available online: https://www.bp.com/content/dam/bp/business-sites/en/global/corporate/pdfs/energy-economics/statistical-review/bp-stats-review-2019-full-report.pdf (accessed on 1 December 2019).
4. Tian, S.; Hamori, S. Time-varying price shock transmission and volatility spillover in foreign exchange, bond, equity, and commodity markets: Evidence from the United States. *N. Am. J. Econ. Financ.* **2016**, *38*, 163–171. [CrossRef]
5. Arouri, M.E.H.; Jouini, J.; Nguyen, D.K. Volatility spillovers between oil prices and stock sector returns: Implications for portfolio management. *J. Int. Money Financ.* **2011**, *30*, 1387–1405. [CrossRef]
6. Soytas, U.; Oran, A. Volatility spillover from world oil spot markets to aggregate and electricity stock index returns in Turkey. *Appl. Energy* **2010**, *88*, 354–360. [CrossRef]

7. Cheung, Y.; Ng, L.K. A causality-in-variance test and its application to financial market prices. *J. Econom.* **1996**, *72*, 33–48. [CrossRef]
8. Arouri, M.E.H.; Lahiani, A.; Nguyen, D.K. Return and volatility transmission between world oil prices and stock markets of the GCC countries. *Econ. Model.* **2011**, *28*, 1815–1825. [CrossRef]
9. Nazlioglus, S.; Erdem, C.; Soytas, U. Volatility spillover between oil and agricultural commodity markets. *Energy Econ.* **2012**, *36*, 658–665. [CrossRef]
10. Nakajima, T.; Hamori, S. Causality-in-Mean and Causality-in-Variance among Electricity Prices, Crude Oil Prices, and Yen-US Dollar Exchange Rates in Japan. *Res. Int. Bus. Financ.* **2012**, *26*, 371–386. [CrossRef]
11. Ewing, B.T.; Malik, F.; Ozfidan, O. Volatility transmission in the oil and natural gas markets. *Energy Econ.* **2002**, *24*, 525–538. [CrossRef]
12. Acaravci, A.; Ozturk, I.; Kandir, S.Y. Natural gas prices and stock prices: Evidence from EU-15 countries. *Econ. Model.* **2012**, *29*, 1646–1654. [CrossRef]
13. Johansen, S.; Juselius, K. Maximum likelihood estimation and inference on cointegration—With applications to the demand for money. *Oxf. Bull. Econ. Stat.* **1990**, *52*, 169–210. [CrossRef]
14. Diebold, F.X.; Yilmaz, K. On the network topology of variance decompositions: Measuring the connectedness of financial firms. *J. Econ.* **2014**, *182*, 119–134. [CrossRef]
15. Diebold, F.X.; Yilmaz, K. *Financial and Macroeconomic Connectedness: A Network Approach to Measurement and Monitoring*, 1st ed.; Oxford University Press: Oxford, UK, 2015.
16. Maghyereh, A.I.; Awartani, B.; Bouri, E. The directional volatility connectedness between crude oil and equity markets: New evidence from implied volatility indexes. *Energy Econ.* **2016**, *57*, 78–93. [CrossRef]
17. Duncan, A.S.; Kabundi, A. Domestic and foreign sources of volatility spillover to South African asset classes. *Econ. Model.* **2013**, *31*, 566–573. [CrossRef]
18. Liow, K.H. Volatility spillover dynamics and relationship across G7 financial markets. *N. Am. J. Econ. Financ.* **2015**, *33*, 328–365. [CrossRef]
19. Sugimoto, K.; Matsuki, T.; Yoshida, Y. The global financial crisis: An analysis of the spillover effects on African stock markets. *Emerg. Mark. Rev.* **2014**, *21*, 201–233. [CrossRef]
20. Toyoshima, Y.; Hamori, S. Measuring the Time-Frequency Dynamics of Return and Volatility Connectedness in Global Crude Oil Markets. *Energies* **2018**, *11*, 2893. [CrossRef]
21. Lovcha, Y.; Perez-Laborda, A. Dynamic frequency connectedness between oil and natural gas volatilities. *Econ. Model.* **2020**, *84*, 181–189. [CrossRef]
22. Ferrer, R.; Shahzad, S.J.H.; Lopez, R.; Jareno, F. Time and frequency dynamics of connectedness between renewable energy stocks and crude oil prices. *Energy Econ.* **2018**, *76*, 1–20. [CrossRef]
23. Zhang, B.; Wang, P. Return and volatility spillovers between china and world oil markets. *Econ. Model.* **2014**, *42*, 413–420. [CrossRef]
24. Oberndorfer, U. Energy prices, volatility, and the stock market: Evidence from the Eurozone. *Energy Policy* **2009**, *37*, 5787–5795. [CrossRef]
25. Kenourgios, D.; Christopoulos, A.G.; Dimitriou, D. Asset Markets Contagion during the Global Financial Crisis. *Multinatl. Financ. J.* **2013**, *17*, 49–76. [CrossRef]
26. Kenourgios, D.; Dimitriou, D. Contagion of the Global Financial Crisis and the real economy: A regional analysis. *Econ. Model.* **2015**, *44*, 283–293. [CrossRef]
27. Baur, D.G. Financial contagion and the real economy. *J. Bank. Financ.* **2012**, *36*, 2680–2692. [CrossRef]
28. Singh, P.; Kumar, B.; Pandey, A. Price and volatility spillovers across North American, European and Asian stock markets. *Int. Rev. Financ. Anal.* **2010**, *19*, 55–64. [CrossRef]
29. Balli, F.; Hajhoj, H.R.; Basher, S.A.; Ghassan, H.B. An analysis of returns and volatility spillovers and their determinants in emerging Asian and Middle Eastern countries. *Int. Rev. Econ. Financ.* **2015**, *39*, 311–325. [CrossRef]
30. Koop, G.; Pesaran, M.H.; Potter, S.M. Impulse response analysis in nonlinear multivariate models. *J. Econ.* **1996**, *74*, 119–147. [CrossRef]
31. Pesaran, H.H.; Shin, Y. Generalized Impulse Response Analysis in Linear Multivariate Models. *Econ. Lett.* **1998**, *58*, 17–29. [CrossRef]
32. Jorion, P. Predicting volatility in the foreign exchange market. *J. Financ.* **1995**, *50*, 507–528. [CrossRef]

33. Blanchard, O.; Simon, J. The long and large decline in U.S. output volatility. *Brook. Pap. Econ. Act.* **2001**, *32*, 135–174. [CrossRef]
34. Tiwari, A.K.; Cunado, J.; Gupta, R.; Wohar, M.E. Volatility spillovers across global asset classes: Evidence from time and frequency domains. *Q. Rev. Econ. Financ.* **2018**, *70*, 194–202. [CrossRef]

© 2020 by the authors. Licensee MDPI, Basel, Switzerland. This article is an open access article distributed under the terms and conditions of the Creative Commons Attribution (CC BY) license (http://creativecommons.org/licenses/by/4.0/).

Article

Connectedness Between Natural Gas Price and BRICS Exchange Rates: Evidence from Time and Frequency Domains

Yijin He [1], Tadahiro Nakajima [1,2] and Shigeyuki Hamori [1,*]

1. Graduate School of Economics, Kobe University, 2-1, Rokkodai, Nada-Ku, Kobe 657-8501, Japan; heheyijin@yahoo.co.jp (Y.H.); nakajima.tadahiro@a4.kepco.co.jp (T.N.)
2. The Kansai Electric Power Company, Incorporated, 6-16, Nakanoshima 3-chome, Kita-Ku, Osaka 530-8270, Japan
* Correspondence: hamori@econ.kobe-u.ac.jp

Received: 27 August 2019; Accepted: 16 October 2019; Published: 18 October 2019

Abstract: In this paper, we investigate the connectedness between natural gas and BRICS (Brazil, Russia, India, China, and South Africa)'s exchange rate in terms of time and frequency. This empirical work is based on the approach of connectedness proposed by Diebold and Yilmaz, who provided an effective way of valuing how much variation in one variable is responsible for the value of other variables, and the method proposed by Baruník and Křehlík, who decomposed the results from Diebold and Yilmaz into different frequencies. We also use the rolling-window method to conduct time-varying analysis. The data used in this paper are from 23 August 2010 to 20 June 2019. We find that the natural gas price hardly influences BRICS's exchange rates, which provides an important practical implication for policymakers, especially in oil-dependent countries.

Keywords: BRICS; exchange rates; connectedness; time domain; frequency domain

1. Introduction

In light of the increasing attention being paid to environmental sustainability, energy systems are gradually transitioning from a dependence on non-renewable resources to the use of environment-friendly resources. This will have a great impact on day-to-day life, economies, businesses, manufacturers, and governments. Compared to coal or petroleum, natural gas has many qualities that makes it burn more efficiently. It also generates fewer emissions of most types of air pollutants, including carbon dioxide. With the expansion of gas pipelines, the increasing number of gas liquefaction plants, and the exploitation of natural gas fields, it is reasonable to consider that the natural gas trade will become more globalized. Natural gas has become a major part of the world's energy consumption, demand, and supply in recent years. In 2018, for example, natural gas consumption rose by 5.3%, one of the fastest rates of growth since 1984. With the continuing rapid expansion in liquefied natural gas (LNG), the inter-regional natural gas trade grew by 4.3%, which was more than double the 10-year average [1]. As reported in the Global Energy Perspective 2019: Reference Case [2], natural gas will be the only fossil fuel whose share of total energy demand continues to increase until 2035, and China will represent nearly half of the global demand growth. Other developing countries are also expected to increase their demand for natural gas.

Brazil, Russia, India, China, and South Africa (BRICS), a group of five fast-growing developing countries, play an important and expanding role in the world economy. In recent years, BRICS have represented an increasing share of global economic growth. According to the International Monetary Fund (IMF), as of 2018, the combined gross domestic product (GDP) of these five nations accounted for 23.2% of the gross world product (GWP). Given the growth of BRICS and the fact that energy is a crucial ingredient for economic development, these countries' relationship with natural gas will only become closer. According to the BP Statistical Review of World Energy [1], in 2018, the total consumption of natural gas in BRICS was 835.8 billion cubic meters (BCM), which accounted for 21.7% of the total global consumption. In terms of imports, China became the second largest importer of LNG, with imports increasing from 4.6 BCM in 2008 to 73.5 BCM in 2018. India was the fourth largest importer, with imports increasing from 11.3 BCM to 30.6 BCM over the same period. In terms of exports, Russia was the largest exporter of pipeline gas. It also accounted for nearly 6% of total LNG exports. As the trade of natural gas is usually settled in US dollars, it is meaningful to study the relationship between the natural gas price and the BRICS's exchange rates.

Against this backdrop, this paper investigates the interdependence between the natural gas price and the BRICS's exchange rate. In doing so, this study is expected to offer valuable insights for market operators, investors, and economists. We use the Henry Hub natural gas futures as the data for the natural gas price. There are two reasons behind this choice of dataset: First, the shale gas revolution in America has dramatically increased US production of shale gas since 2007. World Energy Outlook 2018 [3], produced by the International Energy Agency (IEA), has predicted that natural gas production in America will increase from 976 BCM in 2017 to 1328 BCM in 2040 and that this increase will be mainly due to the growth in shale gas production. Therefore, the Henry Hub natural gas price, which usually represents pricing for the North American natural gas market, has a great influence on the global energy market. We assume that this influence will become stronger over time. The second reason is that there are multiple natural gas price indexes in the world, such as the Japan Korea Marker and the UK National Balancing Point (NBP); however, we cannot predict which price index has strong connectedness with the BRICS's exchange rates. Therefore, we select the Henry Hub price given its characteristics of high liquidity and large trading volume.

Our contribution to the literature is twofold. First, we apply the connectedness methodology from Diebold and Yilmaz [4–6], which allows us to know how pervasive the risk is throughout the entire market by quantifying the contribution of each variable to the system. We also apply the time–frequency version of connectedness proposed by Baruník and Křehlík [7] to find the connectedness between different variables in the short, medium, and long term. Second, to the best of our knowledge, there is not much research on the relationship between the natural gas price and exchange rates. Nevertheless, there are many studies that analyzed the relationship between crude oil prices and foreign exchange rates, and almost all of them show that exchange rates are highly connected to the oil price. For example, in our previous research on the relationship between the West Texas Intermediate (WTI) crude oil price and BRICS's exchange rates using the copula method, we found a significant negative dependence between the two variables. Considering the globalization of natural gas trade, high demand growth (1.6% per year), and the expansive market share in the global energy market (World Energy Outlook 2018 [3] has predicted that, by 2030, natural gas will overtake coal and become the second largest source of energy after oil.), it is reasonable to compare the relationship between the crude oil price and exchange rates with that between the natural gas price and exchange rates. Therefore, in this study, we also aim to determine whether BRICS's exchange rates are closely linked to the natural gas price, as they are to the oil price.

The rest of this paper is organized as follows: A brief review of relevant literature is provided in Section 2. Section 3 introduces the empirical methodology used in this study. Section 4 reports empirical results. Section 5 gives the conclusion. Finally, a robustness analysis is presented in the Appendix A.

2. Literature Review

As we have mentioned above, there is not much literature that has analyzed the relationship between the natural gas price and exchange rates, as far as we know. However, there are many studies on the relationship between the exchange rate and other variables, such as the oil price and the stock market. Chen and Chen [8] investigated the long-term relationship between different crude oil price indexes and G7 countries' exchange rates using the monthly panel data between January 1972 and December 2005. They found that oil prices may account for the movements of the real exchange rate and there is a link between oil prices and real exchange rates. Additionally, from the results of panel predictive regression, they found that the crude oil price has the ability to forecast the future exchange rate. Andrieș et al. [9] identified the patterns of co-movement of the interest rate, stock price, and exchange rate in India using wavelet analysis. They used the data span from July 1997 to December 2010. The empirical results showed that exchange rates, interest rates, and stock prices are linked to each other and that the stock price fluctuations lag behind both the exchange rates and interest rates. Brahm et al. [10] used monthly data to investigate the relationship between the crude oil price and exchange rates in the long term and short term, respectively. The data span was from January 1997 to December 2009. Empirical results indicated exchange rates Granger-caused crude oil prices in the short term, whereas crude oil Granger-caused exchange rates in the long term. Furthermore, based on impulse response analysis, exchange rate shock had a significant negative effect on crude oil prices. Jain and Pratap [11] explored the relationship between global prices of crude oil and gold, the stock market in India, and the USD–INR exchange rate using the DCC-GARCH (dynamic conditional correlation-generalized autoregressive conditional heteroscedasticity) model. They also examined the lead lag linkages among these variables using symmetric and asymmetric non-linear causality tests. They used daily data from the period of 2006 to 2016, finding that a fall in the value of the Indian Rupee and the benchmark stock index was caused by a fall in gold and crude oil prices.

On the empirical side, the methodology used in this paper has already been applied in many fields. Maghyereh et al. [12] used implied volatility indices (VIX) of the daily close price of crude oil in 11 countries. They found that the connectedness between oil and equity was dominated by the transmissions from the oil market to equity markets and most of the linkages between these two markets were established from mid-2009 to mid-2012, a period that witnessed the start of the global recovery. Lundgren et al. [13] studied the renewable energy stock returns and their relation to the uncertainty of currency, oil price, stocks, and US treasury bonds. They used data covering the period from 2004 to 2016, and found that the European stock market depends on renewable energy stock prices. Singh et al. [14] employed a dynamic and directional network connectedness between the implied volatility index (VIX) of the exchange rates of nine major currency pairs and the crude oil using the data between May 2017 and December 2017. They found that crude oil affected currencies more than currencies affected crude oil, but the reverse was true during the crude oil crisis period. Furthermore, their results revealed that EUR–USD is more sensitive to crude oil price fluctuation than others. Ji et al. [15] combined empirical mode decomposition with a connectedness methodology, and examined the dynamic connectedness among crude oil, natural gas, and refinery products using daily data between 3 January 2000 and 15 September 2017. Employing a constant analysis, they found that crude oil and its refinery product tend to be a net transmitter, while the natural gas tends to be a net receiver. In time-varying analysis, they found that the total connectedness generally increased until the 2014 crude oil crash, and then decreased sharply. Lovch and Perez-Laborda [16] used the connectedness method and frequency decomposition method to investigate the relationship between the natural gas and crude oil price during the period from 1994 to 2018. They found that the volatility connectedness varied over time; the connectedness became weak after the financial crisis; and the volatility had long-run effects, except during some specific periods, when volatility shocks transmitted faster but dissipated in the short-run.

3. Empirical Methodology

In this paper, we employ two methods to establish the nature of the relationship between exchange rates and natural gas price. The first method is provided by Diebold and Yilmaz (DY) [4–6], whose approach calculates the connectedness between different objects by introducing variance decomposition into vector autoregression (VAR) models. The second method is based on Baruník and Křehlík (BK) [7], who proposed a new framework to estimate connectedness by using a spectral representation of variance decomposition. In conclusion, the DY framework describes the connectedness as "when shocks are arising in one variable, how would other variables be changing?", whereas the BK framework estimates the connectedness in short-, medium-, and long-term financial cycles.

3.1. Connectedness Table

Based on Diebold and Yilmaz [6], a simplified connectedness table is presented in Table 1, which gives a clear picture of aggregated and disaggregated connectedness.

Table 1. Connectedness table.

	x_1	x_2	...	x_N	From
x_1	d_{11}	d_{12}	...	d_{1N}	$\sum_{j=1}^{N} d_{1j}\ j \neq 1$
x_2	d_{21}	d_{22}	...	d_{2N}	$\sum_{j=1}^{N} d_{2j}\ j \neq 2$
$\vdots$	$\vdots$	$\vdots$	...	$\vdots$	$\vdots$
x_N	d_{N1}	d_{N2}	...	d_{NN}	$\sum_{j=1}^{N} d_{Nj}\ j \neq N$
To	$\sum_{i=1}^{N} d_{i1}\ i \neq 1$	$\sum_{i=1}^{N} d_{i2}\ i \neq 2$	...	$\sum_{i=1}^{N} d_{iN}\ i \neq N$	$\frac{1}{N}\sum_{i,j=1}^{N} d_{ij}\ i \neq j$

Source: Diebold and Yilmaz (2015).

In the table, x_i is the interested variable, whereas d_{ij} is the pairwise directional connectedness from x_j to x_i, which shows what percentage of the h-step-ahead forecast error variance in x_i is due to the shocks in x_j (Equation (1)). We can simply understand d_{ij} as how much future uncertainty of x_i is due to the shocks in x_j:

$$C_{i \leftarrow j} = d_{ij}. \tag{1}$$

The column "From" is the total directional connectedness from x_j to others (Equation (2)), and the row "To" means the total directional connectedness from others to x_i (Equation (3)):

$$C_{. \leftarrow j} = \sum_{\substack{i=1 \\ i \neq j}}^{N} d_{ij}, \tag{2}$$

$$C_{i \leftarrow .} = \sum_{\substack{j=1 \\ j \neq i}}^{N} d_{ij}. \tag{3}$$

We were also interested in net pairwise directional connectedness (Equation (4)) and net total directional connectedness (Equation (5)), which are expressed as a negative value to indicate a net recipient and a positive value to indicate a net transmitter:

$$C_{ij} = C_{j \leftarrow i} - C_{i \leftarrow j}, \tag{4}$$

$$C_i = C_{. \leftarrow i} - C_{i \leftarrow .}. \tag{5}$$

Finally, the total connectedness (Equation (6)), calculated by the grand total of the off-diagonal entries of d_{ij}, is given in the lower-right cell of the connectedness table:

$$C = \frac{1}{N} \sum_{\substack{i,j=1 \\ i \neq j}}^{N} d_{ij}. \tag{6}$$

3.2. Generalized Forecast Error Variance Decomposition (GFEVD)

Diebold and Yilmaz [4] measured connectedness based on forecast error variance decompositions from VAR models, which were introduced by Sims [17] and Koop et al. [18]. However, the calculation of variance decomposition requires orthogonalized shocks and depends on ordering the variables, so Diebold and Yilmaz [5] exploited the generalized forecast error variance decomposition (GFEVD) of Pesaran and Shin [19] to solve those problems. In this paper, we employ the method of GFEVD to calculate the connectedness.

We will give a brief introduction to GFEVD, followed by an explanation of Lütkepohl [20] and Diebold and Yilmaz [6].

For easy understanding, we first consider a VAR (1) process with N-variable:

$$\begin{aligned} y_t &= v + A_1 y_{t-1} + u_t, \quad t = 0, \pm 1, \pm 2 \ldots \\ E(u_t) &= 0 \\ E(u_t u_t') &= \Sigma_u \\ E(u_t u_s') &= 0, \quad t \neq s. \end{aligned} \tag{7}$$

If the generation mechanism starts at time $t = 1$, we get:

$$\begin{aligned} y_1 &= v + A_1 y_0 + u_1 \\ y_2 &= v + A_1 y_1 + u_2 = v + A_1(v + A_1 y_0 + u_1) + u_2 \\ &= (I_N + A_1)v + A_1^2 y_0 + A_1 u_1 + u_2 \\ &\cdots \\ y_t &= (I_N + A_1 + \cdots + A_1^{t-1})v + A_1^t y_0 + \sum_{m=0}^{t-1} A_1^m u_{t-m}\ldots\ldots \end{aligned} \tag{8}$$

If all eigenvalues of A_1 have modulus less than 1 (VAR process is stable), we have:

$$\begin{aligned} (I_N + A_1 + \cdots + A_1^{t-1})v &\to (I_N - A_1)^{-1}v \text{ as } t \to \infty \\ A_1^t y_0 &\to 0 \text{ as } t \to \infty.. \end{aligned} \tag{9}$$

Then, we can rewrite Equation (7) as:

$$\begin{aligned} y_t &= \mu + \sum_{m=0}^{\infty} A_1^m u_{t-m}, \quad t = 0, \pm 1, \pm 2 \ldots \\ \text{where } \mu &\equiv (I_N - A_1)^{-1}v. \end{aligned} \tag{10}$$

Secondly, let us consider a VAR (p) process:

$$y_t = v + A_1 y_{t-1} + \cdots + A_p y_{t-p} + u_t, \quad t = 0, \pm 1, \pm 2, \ldots \tag{11}$$

By using matrices, we can rewrite the VAR (p) process as a VAR (1) process:

$$Y_t = v + AY_{t-1} + U_t$$

$$\underset{(Np*1)}{Y_t} \equiv \begin{bmatrix} y_1 \\ y_2 \\ \cdots \\ y_{t-p+1} \end{bmatrix}, \quad \underset{(Np \times 1)}{v} \equiv \begin{bmatrix} v \\ 0 \\ \cdots \\ 0 \end{bmatrix}$$

$$\underset{(Np \times Np)}{A} \equiv \begin{bmatrix} A_1 & A_2 & \cdots & A_{p-1} & A_p \\ I_N & 0 & \cdots & 0 & 0 \\ 0 & I_N & \cdots & 0 & 0 \\ \cdots & \cdots & \cdots & \cdots & 0 \\ 0 & 0 & \cdots & I_N & 0 \end{bmatrix}, \quad \underset{(Np \times 1)}{U_t} = \begin{bmatrix} u_t \\ 0 \\ \cdots \\ 0 \end{bmatrix}. \quad (12)$$

Similar to Equation (10), Equation (12) can be rewritten as:

$$Y_t = \mu + \sum_{m=0}^{\infty} A^m U_{t-m}, \quad t = 0, \pm 1, \pm 2 \ldots \quad (13)$$

By pre-multiplying a $N \times Np$ matrix $J \equiv [I_N : 0 : \ldots : 0]$, we get:

$$y_t = JY_t = J\mu + \sum_{m=0}^{\infty} JA^m U_{t-m} = J\mu + \sum_{m=0}^{\infty} JA^m J' JU_{t-m}$$
$$= \mu + \sum_{m=0}^{\infty} \Phi_m u_{t-m} \quad (14)$$

$$\underset{(N \times 1)}{\mu} = J\mu, \quad \underset{(N \times N)}{\Phi_m} \equiv JA^m J', \quad \underset{(N \times 1)}{u_t} = JU_t.$$

Finally, we get a moving average (MA) representation of the VAR(p) process:

$$y_t = \mu + \sum_{m=0}^{\infty} \Phi_m u_{t-m}$$
$$E(u_t) = 0$$
$$E(u_t u_t') = \Sigma_u \quad (15)$$
$$E(u_t u_s') = 0, \quad t \neq s.$$

The h-step GFEVD can be expressed as:

$$\omega_{ij,h}^g = \frac{\sigma_{jj}^{-1} \sum_{m=0}^{h-1} (e_i' \Phi_m \Sigma_u e_j)^2}{\sum_{m=0}^{h-1} (e_i' \Phi_m \Sigma_u \Phi_m' e_j)}, \quad (16)$$

where e_i is the i-th column of I_N and σ_{jj} is the j-th diagonal element of Σ_u.

Because the sums of the forecast error variance contribution are not necessarily in agreement, we contribute our generalized connectedness indexes as:

$$d_{ij} = \widetilde{\omega}_{ij}^g = \frac{\omega_{ij,h}^g}{\sum_{j=1}^{N} \omega_{ij,h}^g}. \quad (17)$$

3.3. Spectral Representation of GFEVD

Based on the DY framework, the BK framework defines the general spectral representation of GFEVD and uses it to define the frequency-dependent connectedness measure, which is inspired by the previous research of Geweke [21–23] and Stiassny [24].

Finally, the total connectedness (Equation (6)), calculated by the grand total of the off-diagonal entries of d_{ij}, is given in the lower-right cell of the connectedness table:

$$C = \frac{1}{N} \sum_{\substack{i,j=1 \\ i \neq j}}^{N} d_{ij}. \tag{6}$$

3.2. Generalized Forecast Error Variance Decomposition (GFEVD)

Diebold and Yilmaz [4] measured connectedness based on forecast error variance decompositions from VAR models, which were introduced by Sims [17] and Koop et al. [18]. However, the calculation of variance decomposition requires orthogonalized shocks and depends on ordering the variables, so Diebold and Yilmaz [5] exploited the generalized forecast error variance decomposition (GFEVD) of Pesaran and Shin [19] to solve those problems. In this paper, we employ the method of GFEVD to calculate the connectedness.

We will give a brief introduction to GFEVD, followed by an explanation of Lütkepohl [20] and Diebold and Yilmaz [6].

For easy understanding, we first consider a VAR (1) process with N-variable:

$$\begin{aligned} y_t &= v + A_1 y_{t-1} + u_t, \quad t = 0, \pm 1, \pm 2 \ldots \\ E(u_t) &= 0 \\ E(u_t u_t') &= \Sigma_u \\ E(u_t u_s') &= 0, \quad t \neq s. \end{aligned} \tag{7}$$

If the generation mechanism starts at time $t = 1$, we get:

$$\begin{aligned} y_1 &= v + A_1 y_0 + u_1 \\ y_2 &= v + A_1 y_1 + u_2 = v + A_1(v + A_1 y_0 + u_1) + u_2 \\ &= (I_N + A_1)v + A_1^2 y_0 + A_1 u_1 + u_2 \\ &\cdots \\ y_t &= (I_N + A_1 + \cdots + A_1^{t-1})v + A_1^t y_0 + \sum_{m=0}^{t-1} A_1^m u_{t-m} \cdots \end{aligned} \tag{8}$$

If all eigenvalues of A_1 have modulus less than 1 (VAR process is stable), we have:

$$\begin{aligned} (I_N + A_1 + \cdots + A_1^{t-1})v &\to (I_N - A_1)^{-1} v \text{ as } t \to \infty \\ A_1^t y_0 &\to 0 \text{ as } t \to \infty.. \end{aligned} \tag{9}$$

Then, we can rewrite Equation (7) as:

$$y_t = \mu + \sum_{m=0}^{\infty} A_1^m u_{t-m}, \quad t = 0, \pm 1, \pm 2 \ldots \\ \text{where } \mu \equiv (I_N - A_1)^{-1} v. \tag{10}$$

Secondly, let us consider a VAR (p) process:

$$y_t = v + A_1 y_{t-1} + \cdots + A_p y_{t-p} + u_t, \quad t = 0, \pm 1, \pm 2, \ldots \tag{11}$$

By using matrices, we can rewrite the VAR (p) process as a VAR (1) process:

$$Y_t = v + AY_{t-1} + U_t$$

$$\underset{(Np*1)}{Y_t} \equiv \begin{bmatrix} y_1 \\ y_2 \\ \dots \\ y_{t-p+1} \end{bmatrix}, \quad \underset{(Np \times 1)}{v} \equiv \begin{bmatrix} v \\ 0 \\ \dots \\ 0 \end{bmatrix}$$

$$\underset{(Np \times Np)}{A} \equiv \begin{bmatrix} A_1 & A_2 & \dots & A_{p-1} & A_p \\ I_N & 0 & \dots & 0 & 0 \\ 0 & I_N & \dots & 0 & 0 \\ \dots & \dots & \dots & \dots & 0 \\ 0 & 0 & \dots & I_N & 0 \end{bmatrix}, \quad \underset{(Np \times 1)}{U_t} = \begin{bmatrix} u_t \\ 0 \\ \dots \\ 0 \end{bmatrix}.$$
(12)

Similar to Equation (10), Equation (12) can be rewritten as:

$$Y_t = \mu + \sum_{m=0}^{\infty} A^m U_{t-m}, \quad t = 0, \pm 1, \pm 2 \dots$$ (13)

By pre-multiplying a $N \times Np$ matrix $J \equiv [I_N : 0 : \dots : 0]$, we get:

$$y_t = JY_t = J\mu + \sum_{m=0}^{\infty} JA^m U_{t-m} = J\mu + \sum_{m=0}^{\infty} JA^m J' JU_{t-m}$$
$$= \mu + \sum_{m=0}^{\infty} \Phi_m u_{t-m}$$ (14)

$$\underset{(N \times 1)}{\mu} = J\mu, \quad \underset{(N \times N)}{\Phi_m} \equiv JA^m J', \quad \underset{(N \times 1)}{u_t} = JU_t.$$

Finally, we get a moving average (MA) representation of the VAR(p) process:

$$y_t = \mu + \sum_{m=0}^{\infty} \Phi_m u_{t-m}$$
$$E(u_t) = 0$$
$$E(u_t u_t') = \Sigma_u$$
$$E(u_t u_s') = 0, \quad t \neq s.$$
(15)

The h-step GFEVD can be expressed as:

$$\omega_{ij,h}^g = \frac{\sigma_{jj}^{-1} \sum_{m=0}^{h-1} (e_i' \Phi_m \Sigma_u e_j)^2}{\sum_{m=0}^{h-1} (e_i' \Phi_m \Sigma_u \Phi_m' e_j)},$$ (16)

where e_i is the i-th column of I_N and σ_{jj} is the j-th diagonal element of Σ_u.

Because the sums of the forecast error variance contribution are not necessarily in agreement, we contribute our generalized connectedness indexes as:

$$d_{ij} = \widetilde{\omega_{ij}^g} = \frac{\omega_{ij,h}^g}{\sum_{j=1}^N \omega_{ij,h}^g}.$$ (17)

3.3. Spectral Representation of GFEVD

Based on the DY framework, the BK framework defines the general spectral representation of GFEVD and uses it to define the frequency-dependent connectedness measure, which is inspired by the previous research of Geweke [21–23] and Stiassny [24].

We still consider the MA representation of the VAR(p) process (Equation (15)). The BK framework provides a frequency response function (Equation (18)), which can be obtained as a Fourier transform of the coefficient Φ_m:

$$\Psi(e^{-i\lambda}) = \sum_m e^{-i\lambda m}\Phi_m, \quad i = \sqrt{-1}. \tag{18}$$

The generalized causation spectrum over frequencies $\lambda \in (-\pi, \pi)$ is defined as:

$$(f(\lambda))_{j,k} = \frac{\sigma_{kk}^{-1}\left|(\Psi(e^{-i\lambda})\Sigma_u)_{j,k}\right|^2}{(\Psi(e^{-i\lambda})\Sigma_u\Psi'(e^{+i\lambda}))_{j,j}}, \tag{19}$$

where $(f(\lambda))_{j,k}$ represents the portion of the spectrum of x_j at a given frequency λ due to shocks in x_k. In order to obtain a natural decomposition of variance decomposition to frequencies, a weighting function is defined as:

$$\Gamma_j(\lambda) = \frac{(\Psi(e^{-i\lambda})\Sigma_u\Psi'(e^{-i\lambda}))_{j,j}}{\frac{1}{2\pi}\int_{-\pi}^{\pi}(\Psi(e^{-i\lambda})\Sigma_u\Psi'(e^{-i\lambda}))_{j,j}d\lambda}, \tag{20}$$

where $\Gamma_j(\lambda)$ represents the power of the j-th variable at a given frequency.

The entire range of frequencies' influence of GFEVD from x_j to x_k is expressed as:

$$\omega_{jk}^\infty = \frac{1}{2\pi}\int_{-\pi}^{\pi}\Gamma_j(\lambda)(f(\lambda))_{j,k}d\lambda. \tag{21}$$

Additionally, the GFEVD on specified frequency band $d = (a, b), a, b \in (-\pi, \pi), a < b$, is defined as:

$$\omega_{jk}^d = \frac{1}{2\pi}\int_d \Gamma_j(\lambda)(f(\lambda))_{j,k}d\lambda. \tag{22}$$

As in Section 3.2, we contribute our scaled GFEVD on frequency band d as below, to make sure that the sums of variance contribution are in agreement:

$$d_{ij} = \widetilde{\omega_{jk}^d} = \frac{\omega_{jk}^d}{\sum_k \omega_{jk}^\infty}. \tag{23}$$

4. Empirical Results

4.1. Data

For this study, we collected daily data from Bloomberg, including the Henry Hub natural gas futures price (GASF), and the nominal dollar-denominated exchange rates for the Brazilian Real (BRL), Russian Ruble (RUB), Indian Rupee (INR), offshore Chinese Yuan (CNH), and South African Rand (ZAR). We used the offshore Chines Yuan instead of the onshore Chinese Yuan (CNY) for the reason that China has reformed its exchange rate regime twice, once in 2005 and the other in 2010. Before and after each reform, CNY kept its exchange rate steady for a long time, with almost no fluctuation or only change in a narrow range. Therefore, we chose CNH, which has more fluctuations, to conduct our analysis. In order to match the data availability for CNH, we used the data sample period from 23 August 2010 to 20 June 2019.

The stationary return series were obtained from Equation (18), and are in percentage points:

$$r_{i,t} = 100 \times \ln\left(\frac{p_{i,t}}{p_{i,t-1}}\right). \tag{24}$$

The return series for the natural gas price and exchange rate over time are plotted in Figure 1. Figure 1 shows that the GASF return had the highest volatility compared to the others. We consider that the natural gas price was largely affected by temperature, so most fluctuations occurred concentratedly during the winter season. A small number of fluctuations were recorded in the middle of the year, such as in 2012, when hot weather forecasts and elevated cooling demands created a great demand for natural gas. The RUB return fluctuated drastically at the end of 2014, when the crude oil crash happened, and caused the financial crisis in Russia.

Table 2 provides summary statistics for all return series. CNH has the lowest mean of all return series, as well as standard deviation. Therefore, in some way, CNH remained stable under government regulations. The GASF had the highest standard deviation, as shown in Figure 1. The distribution of all return series significantly deviated from normal, as demonstrated by the Jarque–Bera test.

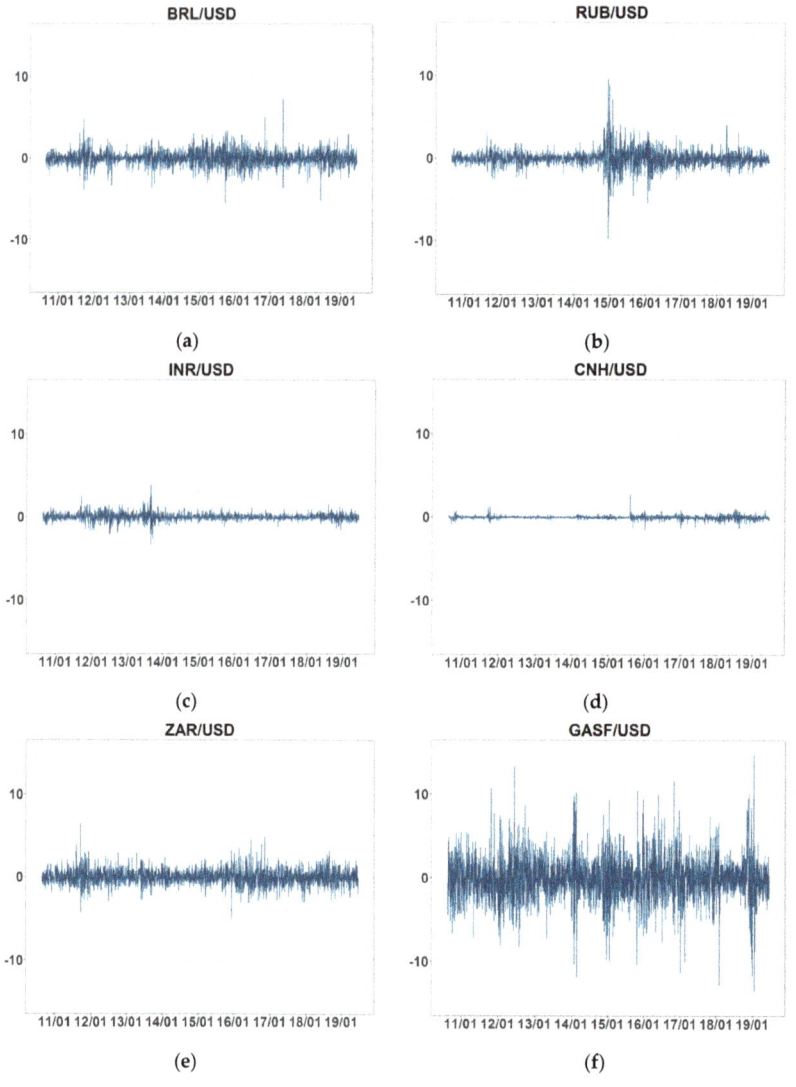

Figure 1. Daily return. Note: BRL, RUB, INR, CNH, and ZAR are Brazilian Real, Russian Ruble, Indian Rupee, offshore Chinese Yuan, and South African Rand, respectively. GASF is the Henry Hub natural gas futures price. (**a**–**f**) refer to BRL, RUB, INR, CNH, ZAR, and GASF return series, respectively.

Table 2. Summary statistics for daily returns.

	Min	Max	Mean	Std Dev	Skewness	Kurtosis	JB-Test
BRL	−5.601	7.270	0.034	0.949	0.140	3.784	1386.046 ***
RUB	−9.771	9.731	0.032	1.024	0.440	13.933	18,741.359 ***
INR	−3.294	3.904	0.017	0.451	0.286	7.546	5509.334 ***
CNH	−1.471	2.747	0.001	0.227	0.473	14.522	20,365.878 ***
ZAR	−5.081	6.444	0.029	0.986	0.271	1.994	411.515 ***
GASF	−18.055	16.691	−0.025	2.759	0.116	4.080	1607.127 ***

Note: BRL, RUB, INR, CNH, and ZAR are Brazilian Real, Russian Ruble, Indian Rupee, offshore Chinese Yuan, and South African Rand, respectively. GASF is the Henry Hub natural gas futures price. The sample period is from 23 August 2010 to 20 June 2019. The JB-Test refers to the Jarque–Bera test for normality. *** indicates rejection of the null hypothesis that the data are normally distributed at the 1% level of significance.

We were interested in not only the connectedness of the return series, but also the volatility connectedness, because volatility can provide a measure of risk and is particularly crisis-sensitive [25]. As volatility is unobserved and must be estimated, we used generalized autoregressive conditional heteroscedasticity (GARCH) models to obtain the volatilities of BRL, INR, CNH, and GASF return series, and Glosten–Jagannathan–Runkle GARCH (GJR-GARCH) models to obtain the volatility of RUB and ZAR return series (for the sake of brevity, the results of the GARCH model and GJR-GARCH model are omitted).

The plots of volatility are presented in Figure 2. For simplicity's sake, we used a different scale for the y-axis in RUB and GASF. The GASF fluctuated violently and most fluctuations accumulated during the winter season, which is consistent with the return series. The volatilities of the five exchange rates reached a high level at the end of 2011, compared to the period before and after, when the eurozone debt crisis reached its peak. The BRL's volatility was turbulent after 2010, especially between 2015 and 2017, when Brazil experienced a severe economic crisis and faced a dramatic economic recession. The volatility of INR reached its peak at the end of 2013, as the Indian rupee had depreciated greatly. The description statistics for volatility are reported in Table 3. Similar to the return series, GASF has the highest standard deviation, whereas CNH has the lowest. All volatilities were skewed and had high kurtosis, indicating that the distributions showed obvious non-normality characteristics. The Jarque–Bera test also verifies our opinion.

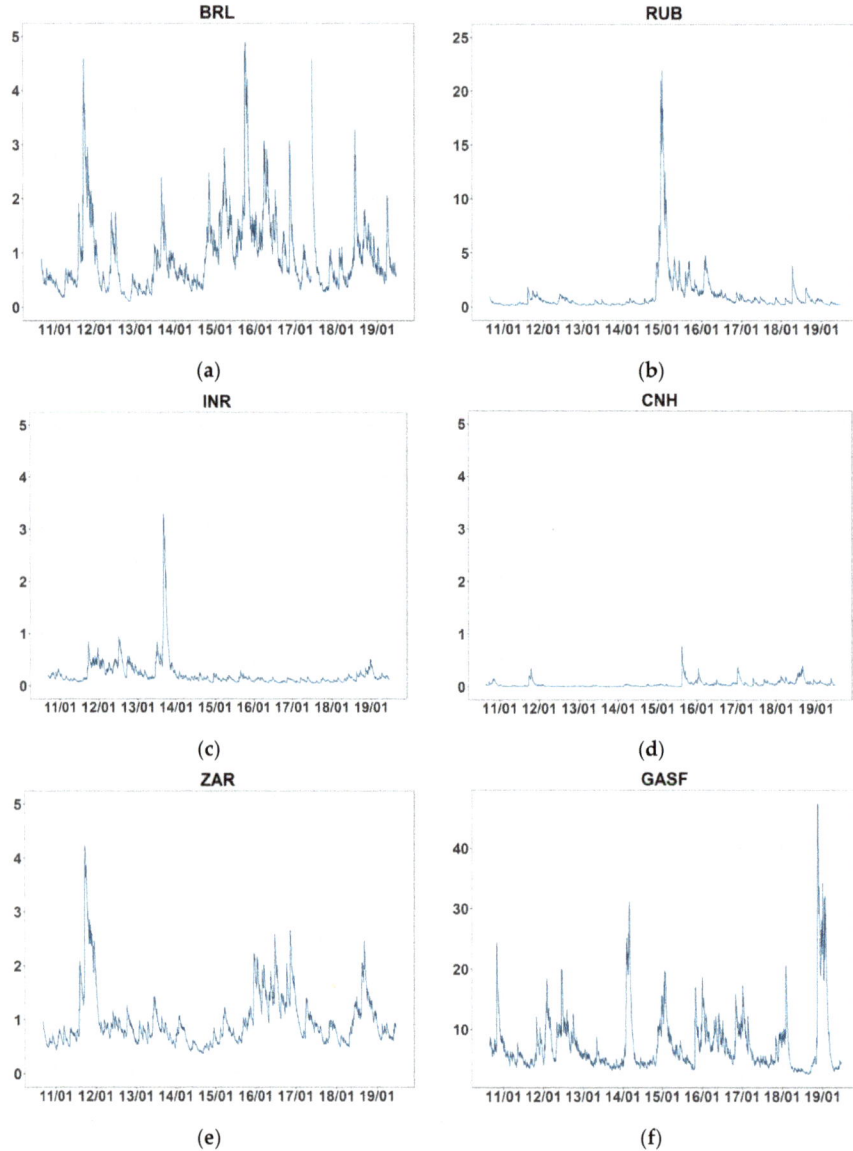

Figure 2. Volatility. Note: BRL, RUB, INR, CNH, and ZAR are Brazilian Real, Russian Ruble, Indian Rupee, offshore Chinese Yuan, and South African Rand, respectively. GASF is the Henry Hub natural gas futures price. (**a**–**f**) refer to the volatility of BRL, RUB, INR, CNH, ZAR, and GASF return series, respectively.

Table 3. Summary statistics for volatilities of daily returns.

	Min	Max	Mean	Std Dev	Skewness	Kurtosis	JB-Test
BRL	0.111	5.466	0.957	0.707	2.127	6.536	5847.958 ***
RUB	0.135	21.907	1.061	2.044	6.108	44.688	206,321.592 ***
INR	0.048	3.297	0.223	0.277	6.275	52.188	276,967.370 ***
CNH	0.005	0.775	0.055	0.068	3.675	21.737	50,619.087 ***
ZAR	0.372	4.234	0.978	0.506	2.205	6.796	6312.786 ***
GASF	2.596	47.407	7.547	5.102	2.837	11.266	15,299.202 ***

Note: BRL, RUB, INR, CNH, and ZAR are Brazilian Real, Russian Ruble, Indian Rupee, offshore Chinese Yuan, and South African Rand, respectively. GASF is the Henry Hub natural gas futures price. The volatilities of BRL, INR, CNY, and GASF return series were calculated by the generalized autoregressive conditional heteroscedasticity (GARCH) (1,1) model, and the volatilities of RUB and ZAR return series were calculated by the Glosten–Jagannathan–Runkle (GJR)-GARCH (1,1) model. The sample period is from 23 August 2010 to 20 June 2019. *** indicates rejection of the null hypothesis that the data are normally distributed at the 1% level of significance.

4.2. Connectedness and Frequency Decomposition

As the calculation of the connectedness index is based on the VAR model, we conducted an augmented Dickey–Fuller (ADF) test for the unit root before applying the data to the VAR model.

However, it is well-known that the unit root hypothesis can be rejected if the data series contain structural break(s) [26–28]. As our sample period is long, from 2010 to 2019, which is almost 9 years, and several big events happened during the sample period, such as the 2014 crude oil crush, which may have had an impact on the economies and caused structural breaks, it is well-founded to consider that structural breaks may exist. Therefore, a Bai–Perron test for structural breaks was conducted. The p-values of the Bai–Perron test for the return series are presented in Table 4. All numbers indicate the acceptance of null hypothesis that no break exists. These results confirm the reliability of the ADF test.

Table 4. Bai–Perron breakpoint test on return series.

	BRL	RUB	INR	CNH	ZAR	GASF
p-value	0.769	0.448	0.316	0.190	0.773	0.536

Note: BRL, RUB, INR, CNH, and ZAR are Brazilian Real, Russian Ruble, Indian Rupee, offshore Chinese Yuan, and South African Rand, respectively. GASF is the Henry Hub natural gas futures price. Each number indicates the p-value of the Bai–Perron breakpoint test.

The results of the ADF test are presented in Table 5. All results show that no unit root exists. The p lags of the VAR model were chosen by the Akaike information criterion (AIC). Return series used the VAR (1) model, whereas volatilities used the VAR (2) model (for the sake of brevity, the results of the VAR model are omitted).

Table 5. Augmented Dickey–Fuller (ADF) test on return series and volatilities.

	BRL	RUB	INR	CNH	ZAR	GASF
			Return			
Dickey–Fuller	−12.050 ***	−11.581 ***	−11.718 ***	−11.470 ***	−14.284 ***	−13.837 ***
			Volatility			
Dickey–Fuller	−5.493 ***	−4.983 ***	−5.471 ***	−6.763 ***	−4.013 ***	−5.422 ***

Note: BRL, RUB, INR, CNH, and ZAR are Brazilian Real, Russian Ruble, Indian Rupee, offshore Chinese Yuan, and South African Rand, respectively. GASF is the Henry Hub natural gas futures price. Each number indicates that Dickey–Fuller is the ADF test statistic. *** indicates rejection of the null hypothesis that a unit root is present in the time series at the 1% level of significance.

The connectedness index based on DY and its spectral representation based on BK of short-, medium-, long-term are reported in Tables 6–9, respectively. The frequency band of short term, medium term, and long term in Table 6 roughly corresponds to 1 day to 5 days, 5 days to 21 days, and more than 21 days, respectively. The diagonal elements in both tables represent the own-market connectedness and are not important in our paper. We have more interest in the off-diagonal elements, which indicate pairwise connectedness between two variables: the values of the To row, which show when one variable receives a shock; how much influence would be exerted on other variables; the values of the "From" column, which measure the composition of one variable's change; and the values of the "Net" row, which reveal whether a variable is a net recipient or a net transmitter. The "GAS-FX" column, which exhibits the net pairwise directional connectedness between GAS and the five exchange rates, is the most critical for our study.

Table 6. Connectedness among the natural gas future price and BRICS's exchange rates.

	BRL	RUB	INR	CNH	ZAR	GASF	From	GAS-FX
				Return				
BRL	69.719	7.790	1.360	3.129	17.692	0.310	30.281	−0.124
RUB	8.335	73.334	2.658	3.303	12.248	0.122	26.666	−0.113
INR	5.502	3.814	77.817	3.686	9.165	0.017	22.183	0.016
CNH	3.609	3.693	3.391	81.305	7.998	0.004	18.695	−0.024
ZAR	16.241	10.551	3.255	6.237	63.673	0.043	36.327	−0.012
GASF	0.433	0.235	0.001	0.028	0.055	99.248	0.752	
To	34.119	26.083	10.665	16.383	47.158	0.496	**22.484**	
Net	3.838	−0.583	−11.517	−2.312	10.831	−0.257		
				Volatility				
BRL	75.206	3.630	0.723	0.092	19.230	1.118	24.794	0.989
RUB	1.040	97.472	0.783	0.060	0.393	0.252	2.528	−3.667
INR	4.205	0.174	91.702	0.446	3.466	0.006	8.298	−0.026
CNH	3.768	0.197	2.853	86.083	6.943	0.156	13.917	−0.107
ZAR	12.265	0.244	1.080	0.449	85.171	0.791	14.829	−0.682
GASF	0.129	3.920	0.032	0.263	1.473	94.183	5.817	
To	21.407	8.165	5.471	1.310	31.506	2.324	**11.697**	
Net	−3.387	5.637	−2.826	−12.607	16.677	−3.493		

Note: BRL, RUB, INR, CNH, and ZAR are Brazilian Real, Russian Ruble, Indian Rupee, offshore Chinese Yuan, and South African Rand, respectively. GASF is the Henry Hub natural gas futures price. From column reports the total directional connectedness from others to x_i. To row reports the total directional connectedness from x_i to others. Net row reports the net total directional connectedness. GAS-FX column reports the net pairwise connectedness between the GASF and exchange rates, which is calculated by the GASF to others minus the others to GASF. The number in red represents the largest value in this system. The number in bold means the total connectedness. All results are expressed as a percentage.

Table 7. Connectedness among the natural gas future price and exchange rates in the frequency domain (short term).

	BRL	RUB	INR	CNH	ZAR	GASF	From	GAS-FX
				Return				
BRL	58.271	6.229	1.130	2.701	14.487	0.282	24.829	−0.111
RUB	6.470	58.659	2.207	2.684	9.513	0.096	20.969	−0.134
INR	3.796	2.569	62.837	2.626	5.975	0.016	14.982	0.015
CNH	2.838	3.122	2.748	65.387	6.316	0.004	15.029	−0.015
ZAR	13.332	8.809	2.767	5.097	51.584	0.042	30.048	0.004
GASF	0.393	0.229	0.001	0.020	0.039	82.464	0.682	
To	26.829	20.958	8.853	13.128	36.330	0.440	**17.756**	
Net	1.999	−0.010	−6.129	−1.901	6.283	−0.242		

Table 7. Cont.

	BRL	RUB	INR	CNH	ZAR	GASF	From	GAS-FX
				Volatility				
BRL	2.783	0.001	0.027	0.000	0.193	0.001	0.222	−0.001
RUB	0.002	1.226	0.011	0.000	0.022	0.000	0.036	−0.010
INR	0.012	0.006	1.107	0.006	0.039	0.001	0.063	−0.002
CNH	0.012	0.000	0.029	5.085	0.057	0.005	0.103	0.003
ZAR	0.085	0.012	0.019	0.004	1.181	0.000	0.120	−0.007
GASF	0.001	0.010	0.004	0.002	0.008	3.005	0.025	
To	0.112	0.030	0.090	0.012	0.319	0.008	**0.095**	
Net	−0.110	−0.006	0.026	−0.091	0.199	−0.017		

Note: BRL, RUB, INR, CNH, and ZAR are Brazilian Real, Russian Ruble, Indian Rupee, offshore Chinese Yuan, and South African Rand, respectively. GASF is the Henry Hub natural gas futures price. From column reports the total directional connectedness from others to x_i. To row reports the total directional connectedness from x_i to others. Net row reports the net total directional connectedness. GAS-FX column reports the net pairwise connectedness between the GASF and exchange rates, which is calculated by the GASF to others minus the others to GASF. The number in red represents the largest value in this system. The number in bold means the total connectedness. The frequency band of short term roughly corresponds to 1 day to 5 days. All results are expressed as a percentage.

Table 8. Connectedness among the natural gas future price and exchange rates in the frequency domain (medium term).

	BRL	RUB	INR	CNH	ZAR	GASF	From	GAS-FX
				Return				
BRL	8.454	1.149	0.170	0.318	2.363	0.020	4.020	−0.009
RUB	1.372	10.802	0.334	0.457	2.011	0.019	4.193	0.015
INR	1.250	0.912	11.041	0.777	2.332	0.001	5.272	0.001
CNH	0.567	0.423	0.474	11.721	1.237	0.000	2.701	−0.006
ZAR	2.148	1.289	0.361	0.840	8.913	0.001	4.639	−0.011
GASF	0.030	0.005	0.000	0.006	0.012	12.391	0.053	
To	5.367	3.777	1.339	2.398	7.956	0.042	**3.480**	
Net	1.347	−0.416	−3.933	−0.303	3.316	−0.012		
				Volatility				
BRL	8.452	0.018	0.091	0.002	0.712	0.006	0.829	−0.002
RUB	0.012	4.530	0.035	0.000	0.082	0.002	0.130	−0.059
INR	0.072	0.034	4.791	0.020	0.193	0.001	0.320	−0.003
CNH	0.075	0.001	0.093	15.124	0.275	0.009	0.452	0.004
ZAR	0.273	0.036	0.062	0.004	3.877	0.001	0.375	−0.032
GASF	0.008	0.061	0.005	0.005	0.033	9.896	0.111	
To	0.440	0.150	0.285	0.029	1.296	0.019	**0.370**	
Net	−0.389	0.020	−0.035	−0.423	0.920	−0.092		

Note: BRL, RUB, INR, CNH, and ZAR are Brazilian Real, Russian Ruble, Indian Rupee, offshore Chinese Yuan, and South African Rand, respectively. GASF is the Henry Hub natural gas future price. From column reports the total directional connectedness from others to x_i. To row reports the total directional connectedness from x_i to others. Net row reports the net total directional connectedness. GAS-FX column reports the net pairwise connectedness between the GASF and exchange rates, which is calculated by the GASF to others minus the others to GASF. The number in red represents the largest value in this system. The number in bold means the total connectedness. The frequency band of medium term roughly corresponds to 5 days to 21 days. All results are expressed as a percentage.

Table 9. Connectedness among the natural gas future price and exchange rates in the frequency domain (long term).

	BRL	RUB	INR	CNH	ZAR	GASF	From	GAS-FX
				Return				
BRL	2.994	0.412	0.060	0.111	0.842	0.007	1.432	−0.003
RUB	0.493	3.873	0.118	0.162	0.724	0.007	1.504	0.006
INR	0.456	0.333	3.940	0.282	0.857	0.000	1.929	0.000
CNH	0.204	0.148	0.169	4.198	0.444	0.000	0.965	−0.002
ZAR	0.761	0.453	0.126	0.299	3.176	0.000	1.640	−0.004
GASF	0.010	0.001	0.000	0.002	0.004	4.392	0.017	
To	1.923	1.347	0.473	0.857	2.872	0.014	**1.248**	
Net	0.491	−0.157	−1.455	−0.108	1.232	−0.003		
				Volatility				
BRL	63.970	3.611	0.606	0.090	18.324	1.111	23.742	0.991
RUB	1.026	91.716	0.737	0.059	0.289	0.250	2.361	−3.598
INR	4.121	0.134	85.804	0.421	3.235	0.003	7.914	−0.020
CNH	3.681	0.196	2.731	65.874	6.611	0.142	13.361	−0.114
ZAR	11.906	0.196	1.000	0.442	80.113	0.790	14.334	−0.642
GASF	0.120	3.848	0.023	0.257	1.432	81.282	5.680	
To	20.855	7.985	5.097	1.269	29.891	2.297	**11.232**	
Net	−2.888	5.624	−2.817	−12.093	15.558	−3.384		

Note: BRL, RUB, INR, CNH, and ZAR are Brazilian Real, Russian Ruble, Indian Rupee, offshore Chinese Yuan, and South African Rand, respectively. GASF is the Henry Hub natural gas future price. From column reports the total directional connectedness from others to x_i. To row reports the total directional connectedness from x_i to others. Net row reports the net total directional connectedness. GAS-FX column reports the net pairwise connectedness between the GASF and exchange rates, which is calculated by the GASF to others minus the others to GASF. The number in red represents the largest value in this system. The number in bold means the total connectedness. The frequency band of long term roughly corresponds to more than 21 days. All results are expressed as percentages.

As shown in Table 6, the total connectedness of the return series is 22.484%, which is almost twice as much as the connectedness between volatilities (11.697%), but both of them are modest. In this system, no matter what the connectedness from the return series or volatilities was, the shocks transmitted from ZAR to BRL contributed the largest value. The BRL was a net transmitter in return connectedness, but a net recipient in volatility connectedness. The RUB was opposite to BRL in that it was a net recipient in the return case, but a net transmitter in the volatility case. Furthermore, INR, CNH, and GASF were net receivers in both cases, whereas ZAR was a net transmitter. By obtaining the absolute value of the "Net" row, we found that INR had the strongest influence (11.517%) in all return series and ZAR was the most powerful variable (16.677%) in volatilities.

In Tables 7–9, we note that the sum of total connectedness in the short term, medium term, and long term is equal to the total connectedness shown in Table 6, which is in agreement with the definition of frequency decomposition for connectedness. It is interesting to find that the total connectedness from the return series is highest in the short term (17.756%), followed by the medium term (3.480%) and long term (1.248%). By contrast, from volatilities, the value is highest in the long term (11.232%), followed by the medium term (0.370%) and short term (0.095%), which means that the uncertainty transmitted by the shock has a long-term impact on the market, rather than the shock itself.

From the values of the GAS-FX column, we found that the net pairwise connectedness between GAS and the exchange rate was higher in volatilities than return series, but both were very weak. All values were almost zero. The possible reason for this is that the GASF data we have chosen are for Henry Hub natural gas, which could be seen as representative of the North American natural gas market. However, as the natural gas pipeline in North America can hardly reach any BRICS countries, and the distance between North America and BRICS countries makes the transportation cost of LNG expensive, whether as an import or export, North American natural gas is not the primary selection for

BRICS countries. Our opinion is also supported by statistics from the BP Statistical Review of World Energy [1]. Whether natural gas is traded by a pipeline or LNG, the quantity being directed from the US, Mexico, and Canada to BRICS countries is very low. Consequently, we could say that the natural gas price is unrelated to the exchange rate in BRICS countries.

4.3. Rolling-Window Analysis

We also conducted a rolling-window analysis to investigate the time-varying connectedness between GAS and exchange rates. The window size was 300 (we also obtained the dynamic connectedness from a window size of 400 and obtained similar results to the result produced from the 300 window size). Figure 3 plots the dynamic total connectedness. From Figure 3, we can see that the total connectedness from the return series begins with a high level (around 40%) in the first few windows, and then falls after 2011 (around 25%), when South Africa joined the BRICS group and the period in which the European debt crisis was at its peak. After a temporary rise in late 2012, the connectedness falls again at the beginning of 2013 (around 20%). From 2013 to mid-2015, the connectedness fluctuates between 20% and 25%, and then rises again to over 30% after mid-2015. The connectedness drops dramatically between 2017 and 2018, from almost 35% to around 20%, and then recovers slowly. The trend of dynamic connectedness from volatilities is similar to that of return series. There are several unusual peaks and troughs in the plot, which we think are related to big events, like the Russian financial crisis (2014), the Brazilian economic recession (2015), and the US–China trade war (2018). Figure 4 presents the frequency decomposition of dynamic connectedness. We find that, whether in the short, medium, or long term, the trend of return connectedness is similar to the dynamic total connectedness. However, for volatilities, in the short and medium term, the connectedness exhibits almost no change (except for some abrupt rises and falls), and the long term has a similar trend to total connectedness. We think that long-term connectedness exerts the most influence in the case of volatility.

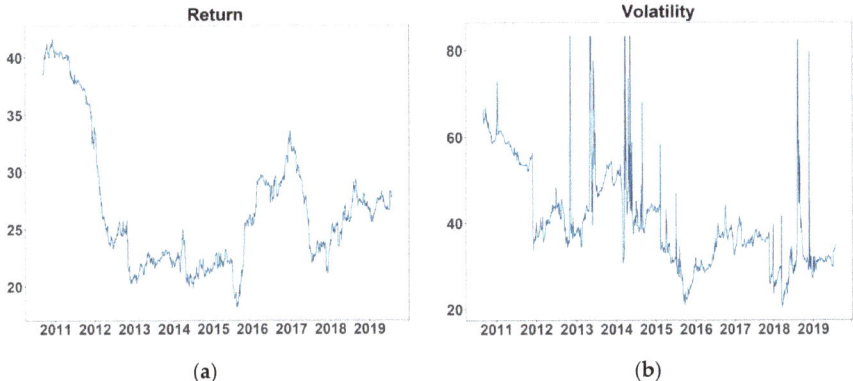

Figure 3. Dynamic connectedness: (**a**) total connectedness of return series and (**b**) total connectedness of volatilities. BRL, RUB, INR, CNH, and ZAR are Brazilian Real, Russian Ruble, Indian Rupee, offshore Chinese Yuan, and South African Rand, respectively.

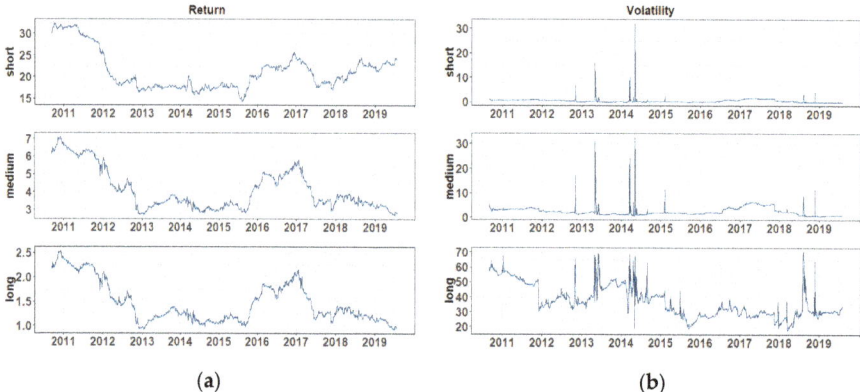

Figure 4. Frequency decomposition of dynamic connectedness: (**a**) frequency decomposition of total connectedness for return series and (**b**) frequency decomposition of total connectedness for volatilities. Note: BRL, RUB, INR, CNH, and ZAR are Brazilian Real, Russian Ruble, Indian Rupee, offshore Chinese Yuan, and South African Rand, respectively. GASF is the Henry Hub natural gas futures price.

The time-varying net pairwise connectedness between GAS and exchange rates return series is plotted in Figure 5. Like the result in Section 4.2, all values were low and almost all of them were below 2.5% in terms of the absolute value, so were negligible. The results from volatilities are presented in Figure 6. There are also several sudden rises and falls, which is consistent with the plot of total connectedness. In the net pairwise connectedness of GAS-BRL, GAS-INR, and GAS-ZAR pairs, except for the abnormal value at some points, the values were almost insignificant; thus, we can hardly say that the GAS has an influence on the exchange rate or vice versa. However, in GAS-RUB and GAS-CNH pairs, there are some significant positive or negative periods during our data span. Before 2014, GAS was a net transmitter to RUB, and then turned into a net recipient after 2014, when Russia was undergoing an economic crisis caused by the oil price crash. After 2016, GAS was a net transmitter to CNH, when Australia became the largest supplier of LNG to China instead of Qatar, and the trade kept increasing after that.

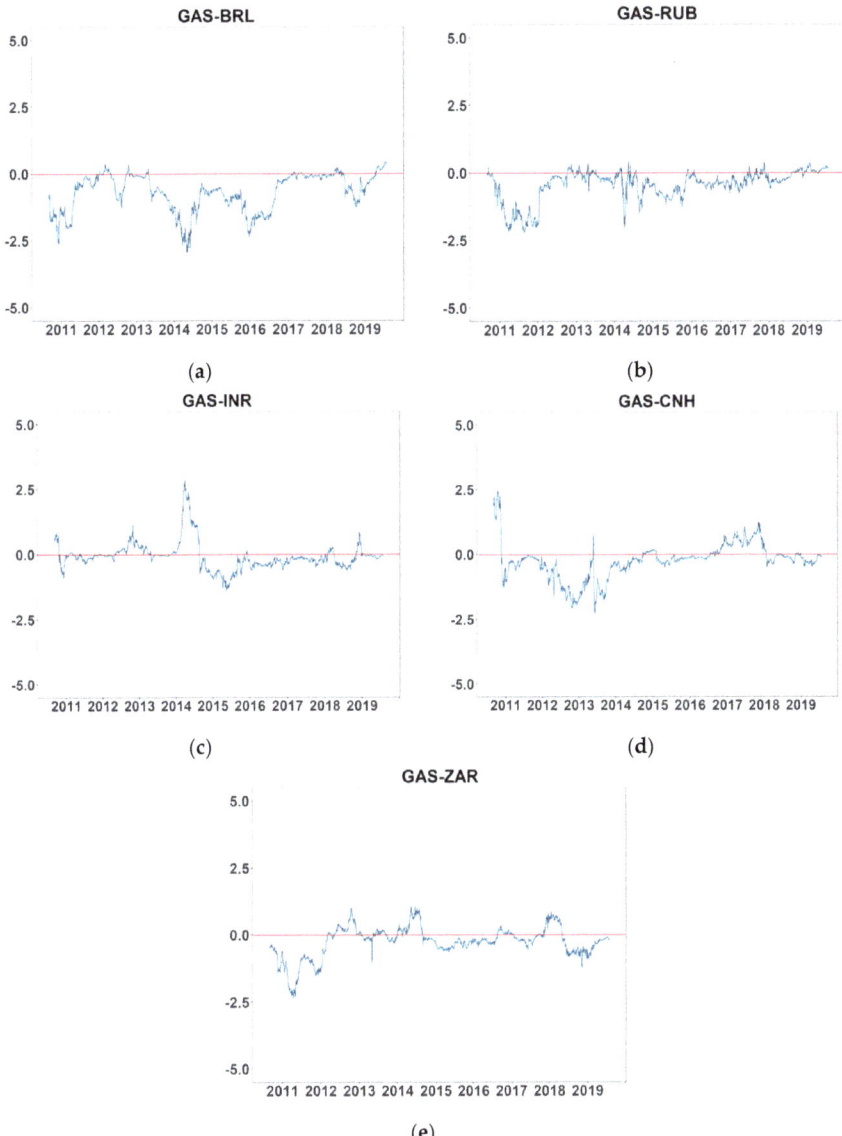

Figure 5. Net pairwise connectedness of return series. (**a–e**) refer to the net pairwise connectedness between GASF and BRL, RUB, INR, CNH, and ZAR return series, respectively. Note: BRL, RUB, INR, CNH, and ZAR are Brazilian Real, Russian Ruble, Indian Rupee, offshore Chinese Yuan, and South African Rand, respectively. GASF is the Henry Hub natural gas futures price.

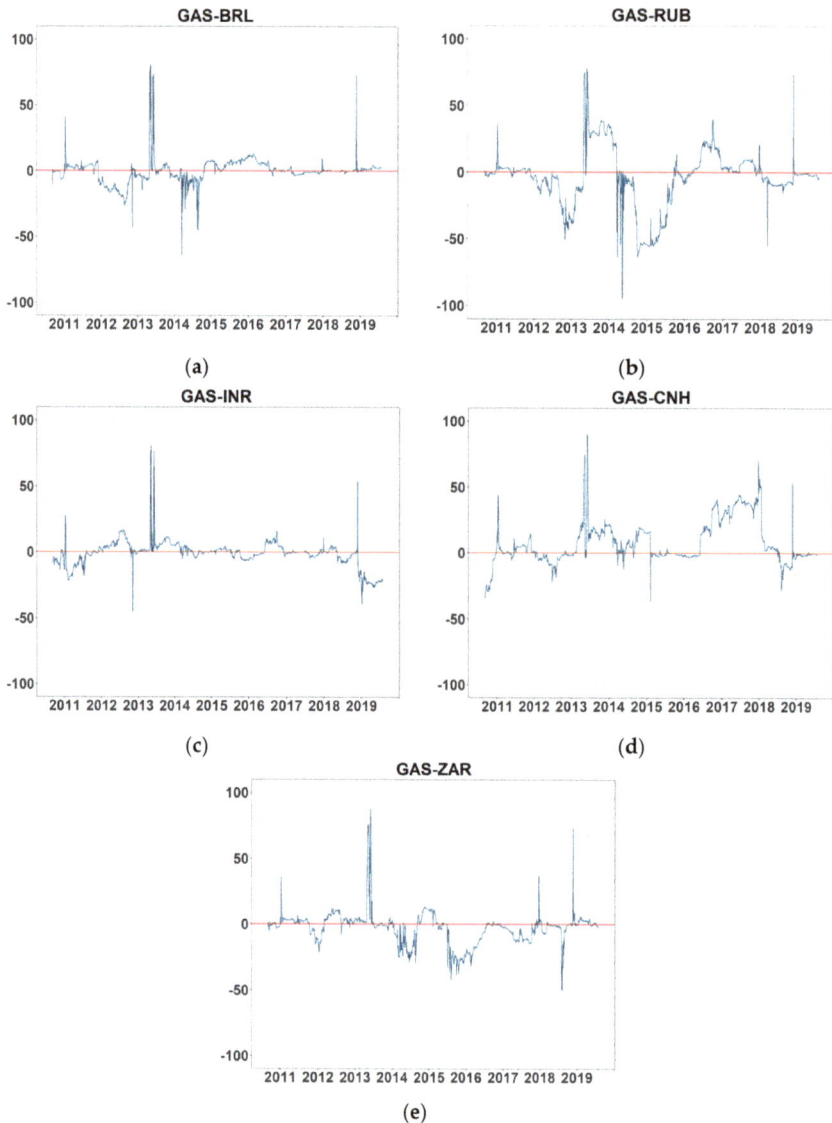

Figure 6. Net pairwise connectedness of volatility. (**a**–**e**) refer to net pairwise connectedness between the volatility of GASF and BRL, RUB, INR, CNH, and ZAR, respectively. Note: BRL, RUB, INR, CNH, and ZAR are Brazilian Real, Russian Ruble, Indian Rupee, offshore Chinese Yuan, and South African Rand, respectively. GASF is the Henry Hub natural gas futures price.

5. Conclusions

This paper examined the connectedness between the Henry Hub natural gas price and the BRICS's exchange rates. To that end, the connectedness methodology from Diebold and Yilmaz (2009, 2012, 2015) as well as frequency decomposition of connectedness proposed by Baruník and Křehlík (2018) were used. We collected data from 23 August 2010 to 20 June 2019 and tested both return series and volatilities from GARCH models.

Our empirical results show that the total connectedness was 22.5% in the return series and 11.7% in volatilities. Compared to results from previous studies—such as Lundgren et al. [13] who found that the total volatility connectedness among renewable energy stock returns, investment assets, and several sources of uncertainty is 67.4%—our results are modest, which means that most variation was due to the variation in the variables themselves. By taking the frequency decomposition of connectedness, we found that, in the return series, the short term contributes to the total connectedness the most, whereas the long term contributes most in relation to volatility. From the results of net pairwise connectedness between the natural gas price and exchange rates, we obtained a value of almost zero in each natural gas and exchange rate pair, which means that natural gas does not play an important role in explaining movements in the exchange rates. We also applied a rolling-window approach to conduct the time-varying analysis. In short, the results are similar to those of the constant analysis and we cannot say for certain that the natural gas price had a great influence on exchange rate movement. Only in the plot of volatility connectedness were there several dramatic fluctuations, which we consider to be connected to some notable events, such as economic crises and trade frictions.

Our results are obviously different from the results of the studies on the relationship between the oil price and exchange rates, such as that conducted by Singh et al. (2018), who found that the total volatility connectedness between the oil price and nine exchange rates reached 72.96%. The shocks transmitted from crude oil to each exchange rate are also significant. We consider some possible reasons for the difference. First, crude oil can be used more widely across different fields than natural gas. For example, it can fuel our cars and make plastics, rubbers, and the like, which are uses that cannot be replaced by natural gas. As indicated in the BP Statistical Review of World Energy [1], crude oil has the highest share in global energy consumption, and its consumption is almost double that of natural gas. Second, the production of crude oil far exceeds that of natural gas. Therefore, whether for energy import countries or energy export countries, crude oil is more easily traded. Third, compared to developed countries, awareness of the environment in developing countries is at a lower level. As the BRICS are the focus of our study, although their consumption of natural gas has increased in recent years, natural gas is still not the primary energy source for these countries (with the exception of Russia). India, China, and South Africa consumed coal the most in 2019, while Brazil consumed oil the most [1].

Although crude oil plays an irreplaceable role in the energy market now, with increasing environmental awareness, we believe that natural gas will become more important and the connectedness between the natural gas price and exchange rates will become stronger in the future.

The empirical evidence in this study may have important implications for policymakers, especially those in oil-dependent countries. As much of the literature shows that exchange rates are highly dependent on the oil price, turbulence in the crude oil market could have a great impact on the foreign currency market, thus causing exchange rate pressure and even economic instability. In order to solve the foreign exchange fluctuation, monetary authorities need to accumulate or reduce foreign exchange reserves, which is not considered desirable in the real world. Changing the dependence structure in relation to energy—from depending on energy that is closely connected with the currency market, such as crude oil, to depending on energy that is hardly connected to the currency market, such as natural gas—could provide an efficient way of maintaining economic stability and reducing exchange rate pressure. By contrast, because of the low connectedness between the natural gas price and exchange rates, foreign exchange fluctuation may barely be affected by the natural gas price. Therefore, for investors, it is less risky to invest in gas-related financial products than oil-related financial products, which are highly connected with currency.

Although this paper conducted thorough research, there were several limitations in the empirical work. First, although we found that natural gas did not have a significant impact on the exchange rate, this result could be influenced by the data selection. We used Henry Hub as our natural gas price data, which represents the North American natural gas market. However, given the restriction of pipelines and high transportation cost, North American countries that produce natural gas may not be the primary selection for BRICS. Second, with technological improvements in exploiting natural gas and the increasing number of gas liquefaction plants, we assumed that the LNG price would have more influence on the exchange rate than the pipeline natural gas price did. However, owing to data limitations, we could only focus on the whole natural gas market, which may be the reason why the connectedness between the natural gas price and exchange rates was modest. Therefore, for further extension of this research, first, we want to collect different natural gas price data, such as the Netherlands Title Transfer Facility (TTF) index and Japan Korea Marker, to exclude the impact of data selection on the results. Second, we want to analyze the relationship between the crude oil price and exchange rates and the relationship between the crude oil price and natural gas price. This would allow us to compare the connectedness between the crude oil price and exchange rates with that between the natural gas price and exchange rates more rationally. Finally, if the data permit, we want to use the data on only LNG to find the connectedness between the natural gas price and foreign exchange rates more precisely.

Author Contributions: Conceptualization, S.H.; investigation, Y.H.; writing—original draft preparation, Y.H.; writing—review and editing, T.N.; project administration, S.H.; funding acquisition, S.H.

Funding: This work was supported by JSPS KAKENHI Grant Number 17H00983.

Acknowledgments: We are grateful to four anonymous referees for their helpful comments and suggestions.

Conflicts of Interest: The authors declare no conflicts of interest.

Appendix A Robustness Analysis

We used the Henry Hub natural gas spot price (GASS) as the natural gas price data to examine the robustness of our results. (The natural gas futures index in the United Kingdom, which is known as the UK National Balancing Point (NBP), was also used to conduct the robustness check, but the results were quite similar to those from GASS, so we only present the connectedness table of NBP and exchange rates (Table A2) in Appendix A).

The plot of GASS's return and volatility series are reported in Figure A1. We found that some values of volatility were extremely large (the maximum is over 800). We think that the reason for this is that the natural gas spot price was more easily affected by the change of demand and supply than the future price, even though the change was small.

We summarize the results of connectedness and the frequency decomposition of short, medium, and long term in Table A1, Table A3, Table A4, Table A5, respectively. The result is quite similar to that of GASF and exchange rates. We also used a 300 rolling-window to conduct the time-varying analysis. The dynamic connectedness and its spectral representation are plotted in Figures A2 and A3, respectively. The net pairwise connectedness of return series and volatilities are illustrated in Figures A4 and A5, respectively. All results are consistent with those from the analysis using the natural gas future price, except for the net pairwise connectedness of return series (Figure A4). Some values are opposite to the result above, but all of them are low, even the maximum value, which is less than 5% and negligible.

The results of robustness confirm the suitability of our proposed approach, which aimed to capture the relationship between the natural gas price and exchange rates.

Table A1. Connectedness between the natural gas spot price and BRICS's exchange rates.

	BRL	RUB	INR	CNH	ZAR	GASS	From	GAS-FX
				Return				
BRL	69.839	7.714	1.456	3.124	17.688	0.179	30.161	−0.249
RUB	8.323	73.405	2.681	3.276	12.230	0.085	26.595	−0.058
INR	5.571	3.833	77.727	3.652	9.066	0.152	22.273	0.040
CNH	3.588	3.648	3.378	81.296	7.953	0.137	18.704	0.136
ZAR	16.233	10.559	3.252	6.206	63.643	0.108	36.357	0.094
GASS	0.429	0.143	0.113	0.001	0.014	99.301	0.699	
To	34.143	25.896	10.879	16.259	46.951	0.660	**22.465**	
Net	3.983	−0.699	−11.394	−2.444	10.593	−0.039		
				Volatility				
BRL	75.533	3.660	0.288	0.354	20.138	0.025	24.467	−0.377
RUB	0.890	97.569	0.750	0.298	0.479	0.014	2.431	−0.006
INR	3.133	0.494	92.791	0.277	3.235	0.070	7.209	−0.234
CNH	3.679	0.213	2.713	85.576	7.123	0.696	14.424	0.678
ZAR	11.251	0.282	0.801	1.067	86.363	0.236	13.637	0.208
GASS	0.402	0.020	0.304	0.018	0.028	99.229	0.771	
To	19.356	4.669	4.856	2.015	31.003	1.041	**10.490**	
Net	−5.111	2.238	−2.354	−12.410	17.367	0.270		

Note: BRL, RUB, INR, CNH, and ZAR are Brazilian Real, Russian Ruble, Indian Rupee, offshore Chinese Yuan, and South African Rand, respectively. GASF is the Henry Hub natural gas futures price. From column reports the total directional connectedness from others to x_i. To row reports the total directional connectedness from x_i to others. Net row reports the net total directional connectedness. GAS-FX column reports the net pairwise connectedness between the GASS and exchange rates, which is calculated by the GASS to others minus the others to GASS. The number in red represents the largest value in this system. The number in bold means the total connectedness. All results are expressed as a percentage.

Table A2. Connectedness between the UK NBP and BRICS's exchange rates.

	BRL	RUB	INR	CNH	ZAR	GASS	From	GAS-FX
				Return				
BRL	69.761	7.791	1.352	3.139	17.707	0.249	30.239	−0.099
RUB	8.337	73.395	2.660	3.307	12.251	0.049	26.605	−0.014
INR	5.472	3.796	77.446	3.686	9.115	0.485	22.554	0.134
CNH	3.600	3.676	3.391	80.904	7.971	0.458	19.096	−0.088
ZAR	16.162	10.491	3.227	6.215	63.346	0.559	36.654	−0.286
GASS	0.348	0.062	0.352	0.546	0.845	97.847	2.153	
To	33.919	25.816	10.983	16.893	47.889	1.800	**22.883**	
Net	3.680	−0.788	−11.571	−2.204	11.236	−0.353		
				Volatility				
BRL	75.164	3.509	0.542	0.277	20.254	0.254	24.836	−0.041
RUB	0.947	96.034	0.952	0.188	0.549	1.329	3.966	1.213
INR	3.494	0.244	90.752	0.328	3.404	1.778	9.248	−1.243
CNH	3.879	0.269	2.684	85.577	6.804	0.788	14.423	0.611
ZAR	11.028	0.318	0.775	0.822	86.380	0.677	13.620	−0.742
GASS	0.295	0.116	3.021	0.177	1.419	94.971	5.029	
To	19.643	4.456	7.974	1.792	32.430	4.827	**11.854**	
Net	−5.193	0.491	−1.274	−12.631	18.810	−0.202		

Note: BRL, RUB, INR, CNH, and ZAR are Brazilian Real, Russian Ruble, Indian Rupee, offshore Chinese Yuan, and South African Rand, respectively. NBP is the UK National Balancing Point. From column reports the total directional connectedness from others to x_i. To row reports the total directional connectedness from x_i to others. Net row reports the net total directional connectedness. GAS-FX column reports the net pairwise connectedness between the NBP and exchange rates, which is calculated by the NBP to others minus the others to NBP. The number in red represents the largest value in this system. The number in bold means the total connectedness. All results are expressed as a percentage.

Table A3. Connectedness between the natural gas spot price and exchange rates in the frequency domain (short term).

	BRL	RUB	INR	CNH	ZAR	GASS	From	GAS-FX
				Return				
BRL	58.953	6.079	1.107	2.772	14.530	0.118	24.606	−0.295
RUB	6.713	58.808	2.160	2.705	9.511	0.057	21.146	−0.060
INR	4.027	2.716	64.161	2.696	6.053	0.150	15.641	0.075
CNH	2.969	3.124	2.809	66.238	6.361	0.096	15.357	0.095
ZAR	13.475	8.616	2.769	5.042	51.620	0.089	29.991	0.081
GASS	0.414	0.118	0.075	0.001	0.008	81.932	0.615	
To	27.597	20.653	8.919	13.216	36.463	0.510	17.893	
Net	2.991	−0.494	−6.722	−2.141	6.472	−0.106		
				Volatility				
BRL	2.698	0.002	0.028	0.002	0.205	0.002	0.238	−0.003
RUB	0.001	1.256	0.011	0.002	0.026	0.001	0.041	−0.003
INR	0.016	0.009	1.099	0.006	0.030	0.000	0.061	−0.001
CNH	0.010	0.002	0.022	5.130	0.056	0.004	0.093	0.001
ZAR	0.090	0.011	0.014	0.006	1.193	0.000	0.121	−0.001
GASS	0.005	0.004	0.001	0.004	0.001	8.637	0.015	
To	0.123	0.028	0.076	0.019	0.318	0.007	**0.095**	
Net	−0.116	−0.014	0.015	−0.075	0.197	−0.008		

Note: The frequency band of short term roughly corresponds to 1 day to 5 days. All results are expressed as a percentage. Note: BRL, RUB, INR, CNH, and ZAR are Brazilian Real, Russian Ruble, Indian Rupee, offshore Chinese Yuan, and South African Rand, respectively. GASF is the Henry Hub natural gas futures price.

Table A4. Connectedness between the natural gas spot price and exchange rates in the frequency domain (medium term).

	BRL	RUB	INR	CNH	ZAR	GASS	From	GAS-FX
				Return				
BRL	8.061	1.197	0.254	0.265	2.325	0.045	4.086	0.032
RUB	1.195	10.750	0.382	0.423	1.997	0.020	4.018	0.001
INR	1.143	0.821	10.068	0.705	2.216	0.003	4.888	−0.025
CNH	0.465	0.392	0.426	11.137	1.177	0.030	2.489	0.030
ZAR	2.045	1.423	0.359	0.856	8.868	0.014	4.697	0.010
GASS	0.013	0.019	0.028	0.000	0.004	13.017	0.065	
To	4.861	3.852	1.448	2.250	7.720	0.113	3.374	
Net	0.775	−0.166	−3.440	−0.240	3.023	0.048		
				Volatility				
BRL	9.181	0.005	0.052	0.020	1.331	0.005	1.413	−0.030
RUB	0.007	3.981	0.043	0.013	0.090	0.002	0.155	−0.003
INR	0.253	0.076	5.828	0.015	0.084	0.002	0.430	−0.012
CNH	0.090	0.003	0.086	14.160	0.385	0.093	0.657	0.087
ZAR	0.486	0.050	0.049	0.005	4.430	0.006	0.596	0.002
GASS	0.036	0.005	0.014	0.007	0.004	35.740	0.065	
To	0.872	0.139	0.244	0.059	1.894	0.108	**0.553**	
Net	−0.541	−0.016	−0.186	−0.598	1.298	0.044		

Note: The frequency band of medium term roughly corresponds to 5 days to 21 days. All results are expressed as a percentage. Note: BRL, RUB, INR, CNH, and ZAR are Brazilian Real, Russian Ruble, Indian Rupee, offshore Chinese Yuan, and South African Rand, respectively. GASF is the Henry Hub natural gas futures price.

Table A5. Connectedness between the natural gas spot price and exchange rates in the frequency domain (long term).

	BRL	RUB	INR	CNH	ZAR	GASS	From	GAS-FX
				Return				
BRL	2.825	0.437	0.095	0.087	0.833	0.016	1.468	0.014
RUB	0.415	3.847	0.138	0.149	0.722	0.007	1.431	0.001
INR	0.401	0.296	3.498	0.250	0.797	0.000	1.744	−0.010
CNH	0.155	0.132	0.144	3.921	0.415	0.010	0.857	0.010
ZAR	0.713	0.520	0.124	0.308	3.155	0.005	1.669	0.003
GASS	0.001	0.006	0.010	0.000	0.002	4.352	0.019	
To	1.685	1.391	0.512	0.793	2.769	0.038	**1.198**	
Net	0.217	−0.040	−1.232	−0.063	1.099	0.019		
				Volatility				
BRL	63.654	3.654	0.208	0.333	18.603	0.018	22.815	−0.343
RUB	0.882	92.332	0.696	0.283	0.362	0.011	2.234	0.001
INR	2.864	0.409	85.864	0.257	3.121	0.068	6.719	−0.221
CNH	3.580	0.209	2.605	66.285	6.682	0.598	13.674	0.591
ZAR	10.675	0.220	0.738	1.056	80.740	0.230	12.919	0.207
GASS	0.361	0.011	0.289	0.008	0.023	54.851	0.691	
To	18.362	4.502	4.535	1.937	28.790	0.926	**9.842**	
Net	−4.454	2.268	−2.183	−11.737	15.872	0.235		

Note: The frequency band of long term roughly corresponds to more than 21 days. All results are expressed as a percentage. Note: BRL, RUB, INR, CNH, and ZAR are Brazilian Real, Russian Ruble, Indian Rupee, offshore Chinese Yuan, and South African Rand, respectively. GASF is the Henry Hub natural gas futures price.

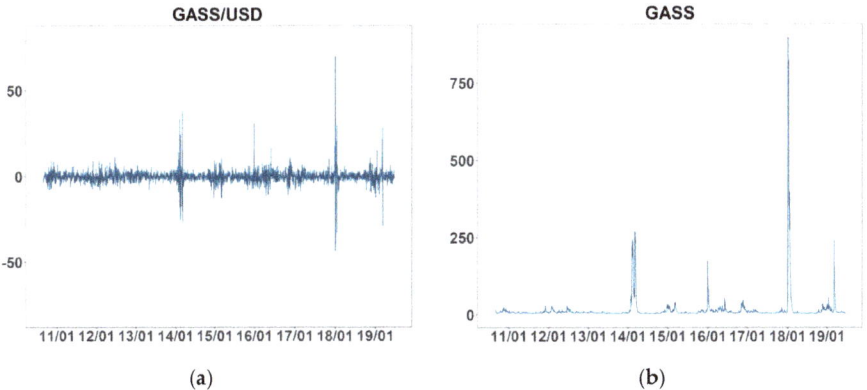

Figure A1. Daily return and volatility: (a) return series and (b) volatility.

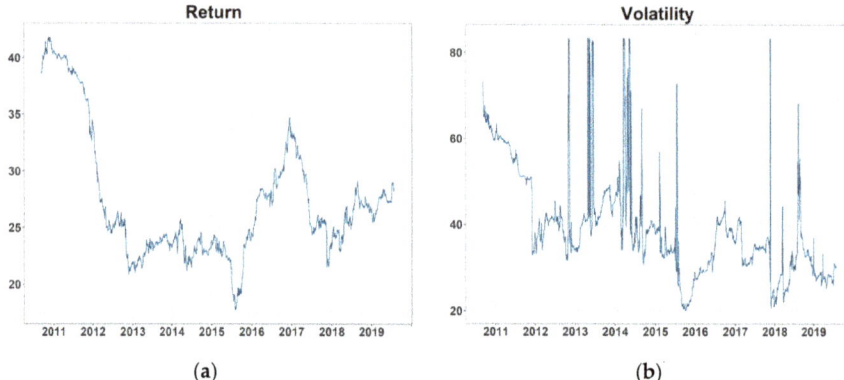

Figure A2. Dynamic connectedness: (**a**) total connectedness of return series and (**b**) total connectedness of volatilities. Note: BRL, RUB, INR, CNH, and ZAR are Brazilian Real, Russian Ruble, Indian Rupee, offshore Chinese Yuan, and South African Rand, respectively. GASF is the Henry Hub natural gas futures price.

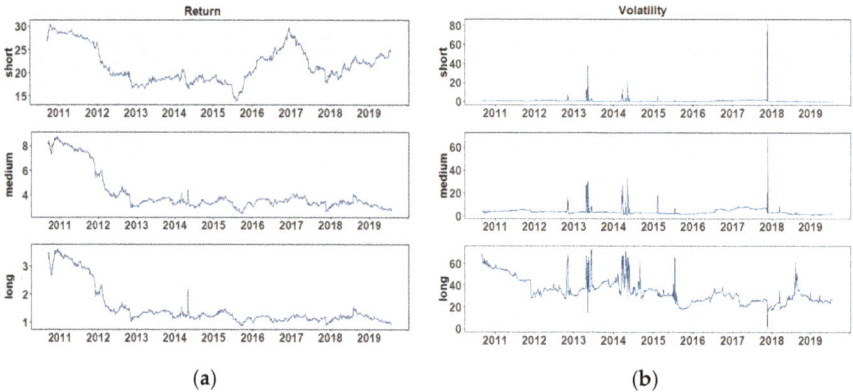

Figure A3. Frequency decomposition of dynamic connectedness: (**a**) frequency decomposition of total connectedness for return series and (**b**) frequency decomposition of total connectedness for volatilities. Note: BRL, RUB, INR, CNH, and ZAR are Brazilian Real, Russian Ruble, Indian Rupee, offshore Chinese Yuan, and South African Rand, respectively. GASF is the Henry Hub natural gas futures price.

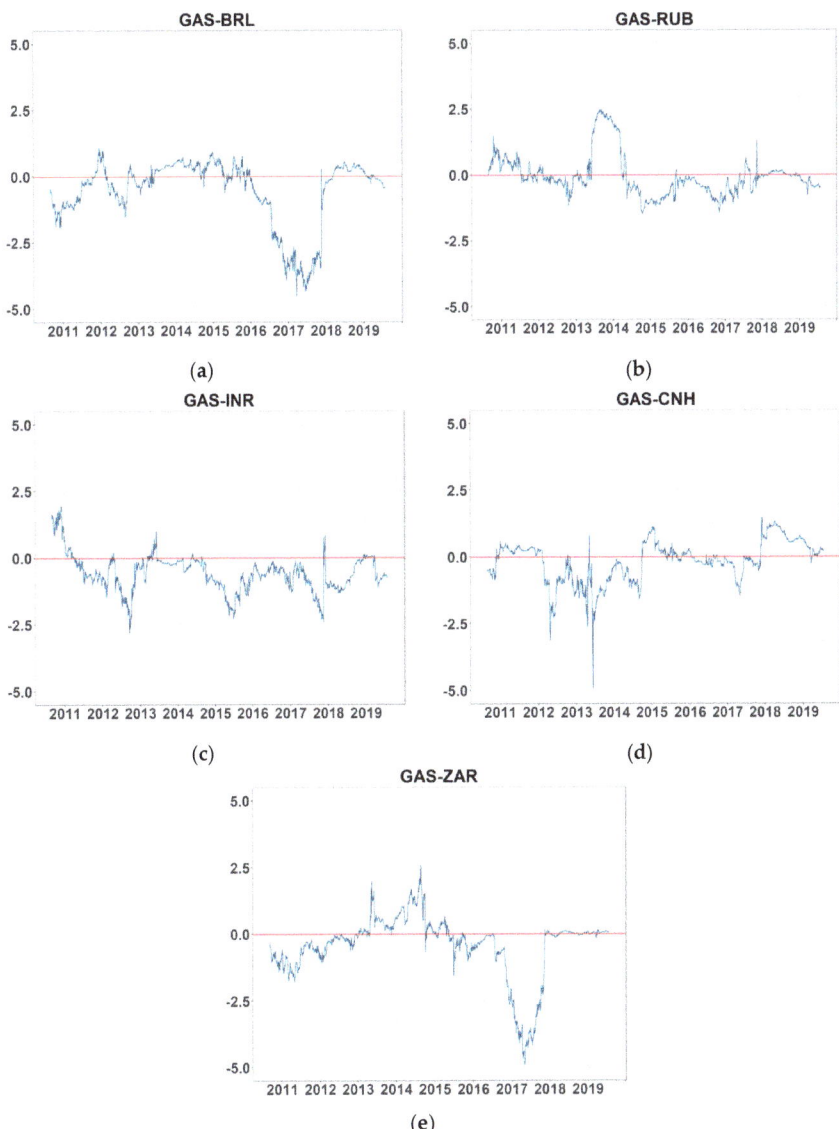

Figure A4. Net pairwise connectedness of return series: (**a**–**e**) refer to the net pairwise connectedness between GASS and BRL, RUB, INR, CNH, and ZAR return series, respectively. Note: BRL, RUB, INR, CNH, and ZAR are Brazilian Real, Russian Ruble, Indian Rupee, offshore Chinese Yuan, and South African Rand, respectively. GASF is the Henry Hub natural gas futures price.

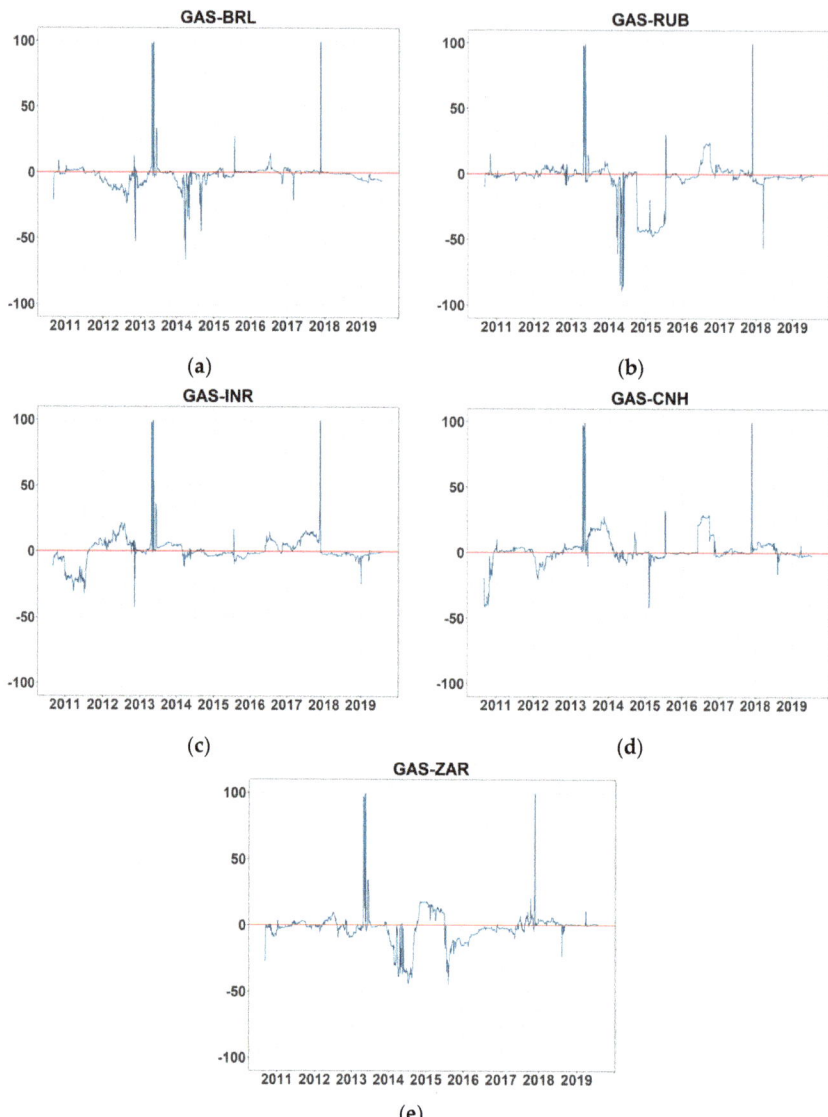

Figure A5. Net pairwise connectedness of volatility. (**a**–**e**) refer to the net pairwise connectedness between the volatility of GASS and BRL, RUB, INR, CNH, and ZAR, respectively. Note: BRL, RUB, INR, CNH, and ZAR are Brazilian Real, Russian Ruble, Indian Rupee, offshore Chinese Yuan, and South African Rand, respectively. GASF is the Henry Hub natural gas futures price.

References

1. BP p.l.c. BP Statistical Review of World Energy 2019. Available online: https://www.bp.com/content/dam/bp/business-sites/en/global/corporate/pdfs/energy-economics/statistical-review/bp-stats-review-2019-full-report.pdf (accessed on 1 July 2019).
2. Mckinsey & Company. Global Energy Perspective 2019: Reference Case. Available online: https://www.mckinsey.com/~{}/media/McKinsey/Industries/Oil%20and%20Gas/Our%20Insights/

Global%20Energy%20Perspective%202019/McKinsey-Energy-Insights-Global-Energy-Perspective-2019_Reference-Case-Summary.ashx (accessed on 22 September 2019).
3. International Energy Agency. World Energy Outlook 2018. 2018. Available online: https://www.iea.org/weo/ (accessed on 22 September 2019).
4. Diebold, F.X.; Yilmaz, K. Measuring Financial Asset Return and Volatility Spillovers, with Application to Global Equity Market. *Econ. J.* **2009**, *119*, 158–171. [CrossRef]
5. Diebold, F.X.; Yilmaz, K. Better to give than to receive: Predictive directional measurement of volatility spillovers. *Int. J Forecast.* **2012**, *28*, 57–66. [CrossRef]
6. Diebold, F.X.; Yilmaz, K. *Financial and Macroeconomic Connectedness: A Network Approach to Measurement and Monitoring*; Oxford University Press: Oxford, UK, 2015.
7. Baruník, J.; Křehlík, T. Measuring the Frequency Dynamics of Financial Connectedness and Systemic Risk. *J. Financ. Econ.* **2018**, *16*, 271–296. [CrossRef]
8. Chen, S.-S.; Chen, H.-C. Oil prices and real exchange rates. *Energy Econ.* **2007**, *29*, 390–404. [CrossRef]
9. Andrieș, A.M.; Ihnatov, I.; Tiwari, A.K. Analyzing time–frequency relationship between interest rate, stock price and exchange rate through continuous wavelet. *Econ. Model.* **2014**, *41*, 227–238. [CrossRef]
10. Brahm, T.; Huang, J.-C.; Sissoko, Y. Crude oil prices and exchange rates: Causality, variance decomposition and impulse response. *Energy Econ.* **2014**, *44*, 407–412.
11. Jain, A.; Pratap, B. Dynamic linkages among oil price, gold price, exchange rate, and stock market in India. *Resour. Policy* **2016**, *49*, 179–185. [CrossRef]
12. Maghyereh, A.I.; Awartani, B.; Bouri, E. The directional volatility connectedness between crude oil and equity markets: New evidence from implied volatility indexes. *Energy Econ.* **2016**, *57*, 78–93. [CrossRef]
13. Lundgren, A.I.; Milicevic, A.; Uddin, G.S.; Kang, S.H. Connectedness network and dependence structure mechanism in green investments. *Energy Econ.* **2018**, *72*, 145–153. [CrossRef]
14. Singh, V.K.; Nishant, S.; Kumar, P. Dynamic and directional network connectedness of crude oil and currencies: Evidence from implied volatility. *Energy Econ.* **2018**, *76*, 48–63. [CrossRef]
15. Ji, Q.; Geng, J.B.; Tiwari, A.K. Information spillovers and connectedness networks in the oil and gas markets. *Energy Econ.* **2018**, *75*, 71–84. [CrossRef]
16. Lovch, Y.; Perez-Laborda, A. Dynamic frequency connectedness between oil and natural gas volatilities. *Econ. Model.* **2019**. [CrossRef]
17. Sims, C.A. Macroeconomics and Reality. *Econometrica* **1980**, *48*, 1–48. [CrossRef]
18. Koop, G.; Pesaran, M.H.; Potter, S.M. Impulse response analysis in nonlinear multivariate models. *J. Econ.* **1996**, *74*, 119–147. [CrossRef]
19. Pesaran, H.H.; Shin, Y. Generalized impulse response analysis in linear multivariate models. *Econ. Lett.* **1998**, *58*, 17–29. [CrossRef]
20. Lütkepohl, H. *New Introduction to Multiple Time Series Analysis*; Springer: Berlin/Heidelberg, Germany, 2005.
21. Geweke, J. Measurement of Linear Dependence and Feedback between Multiple Time Series. *J. Am. Stat. Assoc.* **1982**, *77*, 304–313. [CrossRef]
22. Geweke, J. Measures of Conditional Linear Dependence and Feedback between Time Series. *J. Am. Stat. Assoc.* **1984**, *79*, 907–915. [CrossRef]
23. Geweke, J. The Superneutrality of Money in the United States: An Interpretation of the Evidence. *Econometrica* **1986**, *54*, 1–21. [CrossRef]
24. Stiassny, A. A spectral decomposition for structural VAR models. *Empir. Econ.* **1996**, *21*, 535–555. [CrossRef]
25. Diebold, F.X.; Yilmaz, K. On the network topology of variance decompositions: Measuring the connectedness of financial firms. *J. Econ.* **2011**, *182*, 119–134. [CrossRef]
26. Perron, P. The great crash, the oil price shock, and the unit root hypothesis. *Econometrica* **1989**, *57*, 1361–1401. [CrossRef]

27. Papell, D.H.; Lumsdaine, R.L. Multiple trend breaks and the unit-root hypothesis. *Rev. Econ. Stat.* **1997**, *79*, 212–218.
28. Andrews, D.W.K.; Zivot, E. Further evidence on the great crash, the oil-price shock, and the unit-root hypothesis. *J. Bus. Econ. Stat.* **2002**, *20*, 25–44.

© 2019 by the authors. Licensee MDPI, Basel, Switzerland. This article is an open access article distributed under the terms and conditions of the Creative Commons Attribution (CC BY) license (http://creativecommons.org/licenses/by/4.0/).

Article

Measurement of Connectedness and Frequency Dynamics in Global Natural Gas Markets

Tadahiro Nakajima [1,2,*] and Yuki Toyoshima [3]

1 The Kansai Electric Power Company, Incorporated, Osaka 530-8270, Japan
2 Graduate School of Economics, Kobe University, Kobe 657-8501, Japan
3 Shinsei Bank, Limited, Tokyo 103-8303, Japan; Yuki.Toyoshima@shinseibank.com
* Correspondence: nakajima.tadahiro@a4.kepco.co.jp

Received: 11 September 2019; Accepted: 12 October 2019; Published: 16 October 2019

Abstract: We examine spillovers among the North American, European, and Asia–Pacific natural gas markets based on daily data. We use daily natural gas price indexes from 2 February 2009 to 28 February 2019 for the Henry Hub, National Balancing Point, Title Transfer Facility, and Japan Korea Marker. The results of spillover analyses indicate the total connectedness of the return and volatility series to be 22.9% and 32.8%, respectively. In other words, volatility is more highly integrated than returns. The results of the spectral analyses indicate the spillover effect of the return series can largely be explained by short-term factors, while that of the volatility series can be largely explained by long-term factors. The results of the dynamic analyses with moving window samples do not indicate that global gas market liquidity increases with the increasing spillover index. However, the results identify the spillover effect fluctuation caused by demand and supply.

Keywords: spillover effect; market integration; natural gas market; time frequency dynamics

1. Introduction

The natural gas market has registered an expanding trend, because the fuel transition from coal to natural gas is accelerating in both the industrial and power sectors to reduce greenhouse gas emissions and prevent air pollution, as well as stagnant nuclear power generation. According to BP Statistical Review of World Energy June 2019 [1], the total primary energy consumption in the world grew from 12.4 billion tonnes of oil equivalent (btoe) in 2011 to 13.9 btoe in 2018. Coal consumption was almost constant at 3.6 btoe, while natural gas consumption increased from 2.8 btoe to 3.3 btoe. As a result, the composition ratio of natural gas increased from 22.4% to 23.9%, though the composition ratio of coal decreased from 30.5% to 27.2%. Besides the traditional gas-producing countries in the Middle East and Southeast Asia, others have been increasing their presence as exporting countries, such as Australia (which has been developing large-scale gas wells, including unconventional gas fields) and the United States of America (USA) (which began to export liquefied natural gas (LNG) derived from shale gas). Furthermore, the development of gas fields has been promoted even in Africa recently. Natural gas production skyrocketed from 28 million tonnes of oil equivalent (mtoe), 457 mtoe, and 120 mtoe in 2011 to 112 mtoe, 715 mtoe, and 203 mtoe in 2018, in Australia, the USA, and all of Africa, respectively. On the other hand, consumption is skyrocketing in China and the Middle East, in addition to steady consumption for the members of the Organisation for Economic Co-operation and Development (OECD). Natural gas consumption increased from 24 mtoe, 168 mtoe, and 1142 mtoe in 2011 to 243 mtoe, 476 mtoe, and 1505 mtoe in 2018 in China, all of the Middle East, and the OECD countries.

In the European market, progress in the deregulation of gas business and the decreasing trend of the proportion of long-term contracts linked to crude oil prices has been activating an inter-market arbitrage based on gas pipelines. Moreover, in the global market, the increase in the ratio of LNG to total natural gas trading volumes, the accelerated removal of the destination restriction clause from LNG sales contracts, and the decreasing trend for the percentage of long-term contracts linked to crude oil prices are activating an intercontinental arbitrage based on LNG. As the proportion of spot trading increases, the liquidity of the natural gas market also increases significantly.

According to the Japan Fair Trade Commission of the Government of Japan [2], the natural gas market has not been flexible in terms of region, volume, and price, with the supply chain being considered a special energy market. However, as the trading volume increases with the increase in the number of producing and/or consuming countries, gas market liquidity has also become higher. As a result, the market is changing to a general commodity market.

Natural gas portfolio holders require both security and flexibility in terms of demand and supply because they have to invest heavily and trade globally. Therefore, they need to monitor the relationship between international natural gas markets in terms of revenue and risk management. As such, we adopt the spillover index developed by Diebold and Yilmaz [3] to measure return and volatility spillovers between natural gas price indexes in Europe, North America, and Asia Pacific.

Our expectations for measuring spillover effects between markets are twofold. First, we can easily grasp the potential of a portfolio with a single index obtained by analyzing the relationship of return and volatility between securities that make up that portfolio. We can also overview the potential downside risk of the portfolio without strictly measuring the value at risk and the expected shortfall by a Monte Carlo simulation. Second, we can respond to risks early by using the index as a predictive risk indicator. If we monitor the market that is the source of the spillover, investors can smoothly rebalance their portfolios.

We must test the stationarity of variables because Diebold and Yilmaz [3] develop the spillover index based on the vector moving average (VMA) representation of the vector autoregression (VAR) model. However, the approach proposed by Diebold and Yilmaz [3] is not the Granger–causality test, but just the quantification of the spillover effect. In other words, this technique does not assess whether significant information to predict returns and/or volatility exists, but only estimates how the variables are mutually influential.

However, notwithstanding these limitations of the method, we can obtain not only academic findings concerning the most remarkable transformation market, that is, the natural gas and LNG markets, but also useful information for practitioners, such as regulatory authorities, exchanges, consumers, and suppliers. Further, this index is extremely informative for long-term investment, daily trading, production, and risk management in operating companies that hold such a natural gas portfolio.

In recent years, there have been increased studies on the natural gas market in the context of de-CO_2, the shale gas revolution, and increased liquidity. Kum et al. [4], Das et al. [5], and Bildirici and Bakirtas [6] examine the relationship between natural gas consumption volume and macroeconomic indicators in major developed countries, emerging national economies, and a developing country. Acaravci et al. [7] investigates the relationship between natural gas prices and stock prices in European countries. Nakajima and Hamori [8], Atil et al. [9], Perifanis et al. [10], Tiwari et al. [11], and Xia et al. [12] analyze the relationship between natural gas prices and the other energy prices in the USA. Batten et al. [13] study the relationship between Russian natural gas prices and other energy prices. These previous studies were conducted in the context of causality, spillover, and market integration between natural gas and other economic variables. Moreover, Nakajima [14] argues whether profits can be earned by statistical arbitrage between wholesale electricity futures and natural gas futures.

However, few studies have examined natural gas market integration. Olsen et al. [15], Scarcioffolo and Etienne [16], and Ren et al. [17] discuss the market integration in North America. Nick [18], Osička et al. [19], and Bastianin et al. [20] examine the European natural gas market integration. Shi et al. [21] reveals the interrelationship of LNG prices in Asia. Furthermore, there are few studies that have investigated the natural gas market integration across several regions. Neumann [22] studies the relationship between European and North American markets. Chai et al. [23] study the relationship between the Chinese and the global market. No research has examined the global natural gas market integration, with the exception of that by Silverstovs et al. [24], analyzing the cointegrated relationship between North America, Europe, and Japan based on monthly data from before the shale gas revolution. No studies analyze the spillover effects between North American, European, and Asia-Pacific natural gas markets based on daily data.

We adopt Diebold and Yilmaz's [3] approach to examine spillovers between global natural gas price indexes based on daily data. The correlation coefficient captures only phenomena that do not include the meaning of the relationship. Although the cointegration analysis and the Granger causality test lead to long-term equilibrium and forecast performance, it is difficult to grasp the whole picture at a glance in the case of many variables to be analyzed. Diebold and Yilmaz's [3] approach captures not only pairwise connectedness but also total connectedness. Although Diebold and Yilmaz's [3] approach is not suitable for dynamic descriptions such as multivariate generalized autoregressive conditional heteroscedasticity (GARCH), it is possible to capture dynamic trends with moving window samples. Moreover, we spectrally decompose the Diebold and Yilmaz [3] index by Fourier transform. Baruník and Křehlík [25] utilized the same technique for the first time and later papers applied this approach to the energy market. For instance, Toyoshima and Hamori [26] measure the spillover index for global crude oil markets and decompose it into long-, medium-, and short-term factors. Ji et al. [27] examine the spillover effects between crude oil, heating oil, gasoline, and natural gas in North America and the United Kingdom.

Our contribution to the literature is threefold. The contributions of this paper relate to clarifying the relationships by measuring the connectedness and its frequency dynamics in the global gas market. First, we indicate no progressing integration of the global natural gas market, although we indicate the strong spillover effect between European markets. Moreover, we indicate that the volatility is higher integrated than returns. The Diebold and Yilmaz's [3] approach indicates that the total connectedness of return and volatility is 22.9% and 32.8%, respectively, while each total connectedness is mostly dependent on the pairwise connectedness between European natural gas price indexes. Second, our spectral analyses indicate that long-term factors contribute to volatility spillovers, while short-term factors contribute to return spillovers. We argue that arbitrage might cause short-term return spillovers and long-term memory of volatility might cause long-term volatility spillovers. Finally, our rolling analyses indicate that the above two characteristics continue. Moreover, we argue that regional climate, demand and supply, and incidents might make spillover effects larger or smaller.

The remainder of this paper is organized as follows. Section 2 describes the analyzed data, summary statistics, and preliminary basic analyses. Section 3 explains the adopted methodology. Section 4 presents the empirical results. Section 5 provides a summary of the findings and states conclusions.

2. Data and Preliminary Analyses

2.1. Data

We select natural gas price indicators for the North American, European, and Asia–Pacific markets. We use Henry Hub (HH) futures as the North American index. HH is the name of a distribution hub in Louisiana for the natural gas pipeline system in North America, referring to natural gas delivered at that hub. HH futures are listed on the New York Mercantile Exchange. We adopt National Balancing Point (NBP) and Title Transfer Facility (TTF) futures as the European index. NBP futures are futures of natural gas at a virtual trading location in the United Kingdom, being listed on the Intercontinental Exchange (ICE). This market is related to production in the North Sea gas field, which is trading with the European continent by pipelines, and consumption within the United Kingdom. TTF futures are the futures of natural gas at a virtual trading point in the Netherlands, being listed on the ICE. This market reflects the continental European market, which consists of trading by long-haul pipelines and LNG. We utilize the Japan/Korea Marker (JKM) as the Asia-Pacific index, which is associated with the short-term trading market for LNG in the Asia–Pacific region. It is provided by Platts, which assesses benchmark prices in physical energy markets. We obtain the daily data from February 2, 2009 to February 28, 2019 from Bloomberg.

Figure 1 shows the time series for these indexes. First, HH fluctuates independently from the other variables. Second, the two European indexes move together. Finally, JKM is linked to the European market over a specific period. As a result of the Great East Japan Earthquake on 11 March 2011, the amount of power generated by nuclear power plants has decreased significantly in Japan. Thus, we can assume that the Asia-Pacific LNG market was tight, and therefore the JKM was soaring. Moreover, these European indexes might have increased as a result of arbitrage trading with the Asia-Pacific LNG market. HH's downtrend around 2012 might be caused by the expectations of increased shale gas production, while the price spike in 2014 was caused by the North American cold wave.

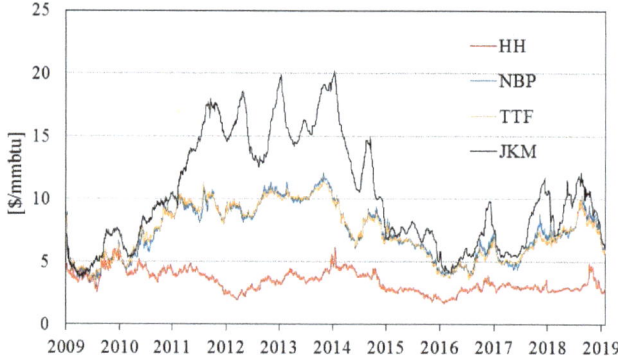

Figure 1. Time series plots of natural gas price. (HH = Henry Hub, NBP = National Balancing Point, TTF = Title Transfer Facility, JKM = Japan/Korea Marker).

Table 1 presents the summary statistics of the return and volatility series of each index. There are 2436 observations in each case. The JKM return series has a distribution biased to the right and the other series have distributions biased to the left, as skewness is negative only for the JKM return series. The kurtosis of all series are extremely large. In other words, all series have fat tail distributions. We can reject the hypothesis that each series is normally distributed by the Jarque-Bera statistics calculated from the skewness and kurtosis of each series.

Table 1. Summary statistics.

Statistic	Return				Volatility *			
	Henry Hub (HH)	National Balancing Point (NBP)	Title Transfer Facility (TTF)	Japan/Korea Marker (JKM)	HH	NBP	TTF	JKM
Observations	2436	2436	2436	2436	2436	2436	2436	2436
Mean	0.0%	0.0%	0.0%	0.0%	2.9%	2.3%	2.0%	1.3%
Median	−0.1%	−0.1%	−0.1%	−0.0%	2.7%	2.1%	1.8%	1.0%
Maximum value	30.7%	42.9%	29.6%	15.4%	8.7%	11.2%	7.9%	10.4%
Minimum value	−16.5%	−12.3%	−12.0%	−22.6%	1.3%	0.7%	0.7%	0.3%
Standard deviation	3.2%	2.7%	2.3%	1.7%	1.1%	1.0%	0.9%	0.8%
Skewness	0.92	2.3	1.5	−0.9	1.5	1.6	1.3	3.0
Kurtosis	10.0	35.0	21.0	31.3	6.5	9.0	5.6	21.2
Jarque–Bera	5358	106,011	33,912	81,799	2182	4654	1353	37,478
(p-value)	(0)	(0)	(0)	(0)	(0)	(0)	(0)	(0)

*: calculated by the stochastic volatility (SV) model.

As described in Toyoshima and Hamori [26], let x_t be a return with mean zero and variance $\exp(h_t)$. The SV model can be expressed as follows:

$$x_t|h_t \sim N(0, exp(h_t))$$

$$h_t|h_{t-1}, \mu, \varphi, \sigma_\tau \sim N(\mu + \varphi(h_{t-1} - \mu), \sigma_\tau^2)$$

$$h_0|\mu, \varphi, \sigma_\tau \sim N(\mu, \sigma_\tau^2/(1-\varphi^2))$$

where μ, φ, and σ_τ are the level of log variance, the persistence of log variance, and the volatility of log variance, respectively.

2.2. Preliminary Analyses

The condition for the VMA representation of the VAR model is that all variables are stationary. Accordingly, we test for the stationarity status of all series by the augmented Dickey–Fuller (ADF) test. Table 2 presents the results. The ADF test rejects the null hypothesis that all variables have a unit root. Therefore, we can utilize the VMA representation.

We estimate the coefficient of the diagonal Baba, Engle, Kraft, and Kroner (BEKK) GARCH model to calculate the dynamic correlation coefficients. The BEKK model is one of several variations of multivariate GARCH models depending on the formulation of the time-varying variance–covariance matrix. The BEKK model was developed by Engle and Kroner [28], and it guarantees a positive estimated variance. Moreover, the BEKK model can dynamically calculate the correlation coefficient. When there are k variables, the variance-covariance matrix H_t of the BEKK model is as follows:

$$H_t = L'L + M'\varepsilon_{t-1}\varepsilon'_{t-1}M + N'H_{t-1}N \quad (1)$$

where M and N are a $k \times k$ matrix, and L is a $k \times k$ symmetric matrix.

Table 2. Unit root tests.

Series	Variables	Exogenous	ADF–t Value (p-Value)
Return	Henry Hub (HH)	Constant	−54.12 (0.000)
		Constant, Trend	−54.11 (0.000)
	National Balancing Point (NBP)	Constant	−48.85 (0.000)
		Constant, Trend	−48.84 (0.000)
	Title Transfer Facility (TTF)	Constant	−47.65 (0.000)
		Constant, Trend	−47.64 (0.000)
	Japan/Korea marker (JKM)	Constant	−19.70 (0.000)
		Constant, Trend	−19.73 (0.000)
Volatility	HH	Constant	−5.13 (0.000)
		Constant, Trend	−5.31 (0.000)
	NBP	Constant	−7.18 (0.000)
		Constant, Trend	−7.26 (0.000)
	TTF	Constant	−6.29 (0.000)
		Constant, Trend	−6.37 (0.000)
	JKM	Constant	−16.60 (0.000)
		Constant, Trend	−16.85 (0.000)

We calculate the correlation coefficients over the entire period, namely from 2 February 2009 to 28 February 2019, to understand the simultaneous relationship between variables as a phenomenon. Tables 3 and 4 present the correlation coefficients for the return and volatility series, respectively. The correlation coefficient between the NBP and TTF is the largest for both the return and volatility series. Furthermore, the correlation coefficients of the volatility series are larger than those of the return series under each combination of variables. This fact indirectly indicates the high sustainability of volatility.

Table 3. Correlation coefficient of return series.

Variables	Henry Hub (HH)	National Balancing Point (NBP)	Title Transfer Facility (TTF)	Japan/Korea Marker (JKM)
HH	1.000			
NBP	0.064	1.000		
TTF	0.089	0.798	1.000	
JKM	0.047	0.102	0.095	1.000

Table 4. Correlation coefficient of volatility series.

Variables	Henry Hub (HH)	National Balancing Point (NBP)	Title Transfer Facility (TTF)	Japan/Korea Marker (JKM)
HH	1.000			
NBP	0.397	1.000		
TTF	0.465	0.940	1.000	
JKM	0.223	0.402	0.465	1.000

Figure 2 provides the time series of the dynamic correlation coefficients between each pair of return series. Although the NBP and TTF have the highest correlation coefficient in Table 4, even that dynamic correlation coefficient temporarily decreases to 0. Conversely, even if a correlation coefficient over the entire period is low, the dynamic correlation coefficient may temporarily increase.

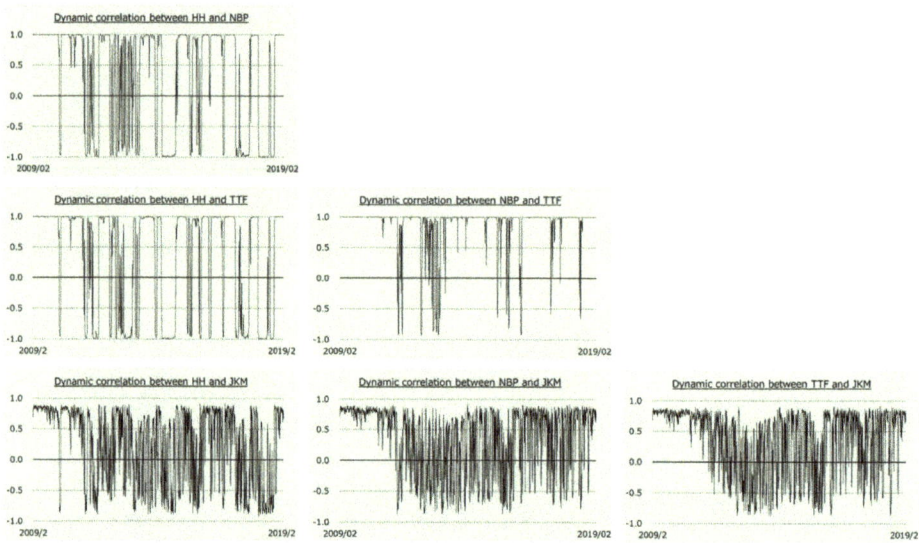

Figure 2. Dynamic correlation between each return series.

Figure 3 provides the time series of the dynamic correlation coefficients between each volatility series. We find that the correlation and inverse correlation phases are clear because volatility accumulates at a high rate. However, only the dynamic correlation coefficient between the NBP and TTF continues to be almost +1.

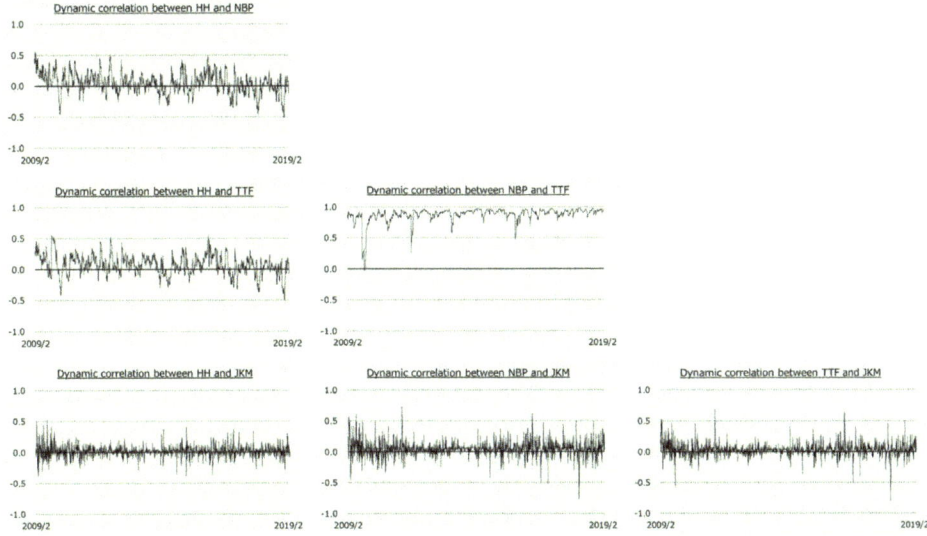

Figure 3. Dynamic correlation between each volatility series.

3. Methodology

3.1. Spillover Index

A correlation coefficient only represents a simultaneous phenomenon between variables. Therefore, our analysis on the relationship between variables needs to go further. We begin by calculating the index proposed by Diebold and Yilmaz [3] which can capture the spillover effect between variables.

We assume the following covariance stationary four-variable VAR (p):

$$x_t = \sum_{l=1}^{p} \Phi_l x_{t-l} + \varepsilon_t, \tag{2}$$

where x_t is the four-dimensional vector of return or volatility, which must be a stationary series; Φ_l are the 4×4 coefficient matrices; p is the lag length based on the Schwarz information criterion; and ε_t is an independently and identically distributed sequence of four-dimensional random vectors with zero mean and covariance matrix $E(\varepsilon_t \varepsilon_t') = \Sigma$.

We can represent the above VAR model in the following VMA:

$$x_t = \sum_{l=0}^{\infty} A_l \varepsilon_{t-l} \tag{3}$$

where $A_l = \sum_{l=1}^{p} \Phi_l A_{t-l}$, A_0 being a 4×4 identity matrix with $A_l = 0$ for $l < 0$.

We define the spillover effect from the mth to the lth market up to H-step-ahead as the following equation by using the H-step-ahead forecast error variance decompositions:

$$\theta_{lm} = \frac{\sigma_{mm}^{-1} \sum_{h=0}^{H-1} (e_l' A_h \Sigma e_m)^2}{\sum_{h=0}^{H-1} e_l' A_h \Sigma A_h' e_l} \tag{4}$$

where σ_{mm} is the standard deviation of the error term for the mth equation and e_l is the selection vector, with one as the lth element and zeros otherwise.

We normalize each entry of the variance decomposition matrix by the row sum, that is, 4, as the pairwise connectedness:

$$\widetilde{\theta_{lm}} = \frac{\theta_{lm}}{\sum_{m=1}^{4} \theta_{lm}} = \frac{\theta_{lm}}{4} \tag{5}$$

We define the sum of pairwise connectedness as total connectedness:

$$S = \frac{\sum_{l=1}^{4} \sum_{m=1, l \neq m}^{4} \widetilde{\theta_{lm}}}{4} \tag{6}$$

The numerator is the sum of the spillover effects, excluding the spillover effects on itself. In other words, the total connectedness means the sum of the relative proportion of the portfolio's response to a shock.

Moreover, we measure the directional spillover effects received by the lth market from all other markets as:

$$S_{l\cdot} = \frac{\sum_{l=1, l \neq m}^{4} \widetilde{\theta_{lm}}}{4} \tag{7}$$

Similarly, we measure the directional spillover effects transmitted by the lth market to all other markets as:

$$S_{\cdot l} = \frac{\sum_{l=1, l \neq m}^{4} \widetilde{\theta_{ml}}}{4} \tag{8}$$

3.2. Spectral Analysis

Based on the Fourier transform utilized by Baruník and Křehlík [25], we spectrally decompose Diebold and Yilmaz's [3] indexes into short-term (1 to 5 business days) factors, medium-term (6 to 20 business days) factors, and long-term (from 21 business days onward) factors.

The Fourier transform of Equation (2) is as follows:

$$f(\omega)_{lm} = \frac{\sigma_{mm}^{-1}\left(\left(\sum_{h=0}^{H-1} e^{-i\omega h} A_h\right)\Sigma\right)_{lm}^{2}}{\left(\left(\sum_{h=0}^{H-1} e^{-i\omega h} A_h\right)\Sigma\left(\sum_{h=0}^{H-1} e^{i\omega h} A_h\right)'\right)_{ll}} \tag{9}$$

This is the spillover effect from the mth market to the lth market up to H-step-ahead, which is expressed by angular frequency ω.

We define the weighting function, the ratio of ω component to all-frequency components concerning the spillover to the lth variable, as follows:

$$\Gamma_l(\omega) = \frac{\left(\left(\sum_{h=0}^{H-1} e^{-i\omega h} A_h\right)\Sigma\left(\sum_{h=0}^{H-1} e^{i\omega h} A_h\right)'\right)_{ll}}{\frac{1}{2\pi}\int_{-\pi}^{\pi}\left(\left(\sum_{h=0}^{H-1} e^{-i\lambda h} A_h\right)\Sigma\left(\sum_{h=0}^{H-1} e^{i\lambda h} A_h\right)'\right)_{ll} d\lambda} \tag{10}$$

The spillover index from the mth market to the lth market in all bands is expressed as:

$$(\theta_\infty)_{lm} = \frac{1}{2\pi}\int_{-\pi}^{\pi} \Gamma_l(\omega) f(\omega)_{lm} d\omega \tag{11}$$

The spillover index from the mth market to the lth market in band d is expressed as:

$$(\theta_d)_{lm} = \frac{1}{2\pi}\int_d \Gamma_l(\omega) f(\omega)_{lm} d\omega \tag{12}$$

We convert $(\theta_d)_{lm}$ to the relative contribution $\left(\widetilde{\theta_d}\right)_{lm}$ as:

$$\left(\widetilde{\theta}_d\right)_{lm} = \frac{(\theta_d)_{lm}}{\sum_{l=1}^{4}(\theta_\infty)_{lm}} \tag{13}$$

We calculate the total spillover index in the d band as:

$$C_d = \frac{\sum_{l=1}^{4}\sum_{m=1,l\neq m}^{4}\left(\widetilde{\theta}_d\right)_{lm}}{\sum_{l=1}^{4}\sum_{m=1}^{4}\left(\widetilde{\theta}_d\right)_{lm}} \tag{14}$$

Naturally, this is consistent with Diebold and Yilmaz's [3] spillover index.

3.3. Rolling Analysis

It is insufficient to focus on static spillover indicators, which are calculated by the Diebold and Yilmaz [3] and Baruník and Křehlík [25] for the entire period. To capture the dynamics of the spillover effects, we employ the rolling window approach as a sampling method. As shown in Figure 4, this study fixes the moving window sample size to 300 trading days and offsets the window by one business days every time we perform an analysis.

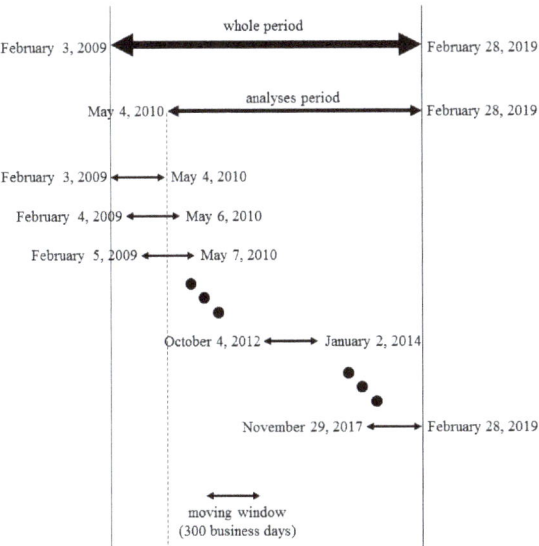

Figure 4. Illustration of moving window procedure.

4. Empirical Results

4.1. Spillover Index

The spillover analyses results of the return series are reported in Table 5. The spillover index from HH to the other variables and from the other variables to HH are 0.42% and 0.57%, respectively. HH fluctuates independently of the other variables.

The spillover indexes from NBP to TTF and from TTF to NBP are both above 40%. While the total connectedness is 22.9%, the spillover indexes from the other variables to TTF and from TTF to the other variables are 10.7% and 10.8%, respectively. Moreover, the spillover indexes from the other variables to NBP and from NBP to the other variables are 10.5% and 11.1%, respectively. TTF and NBP have a high presence as both the sources and destinations of spillover effects, because these two natural gas markets in Europe are almost integrated, although HH and JKM fluctuate relatively independently. However, only the spillover index from NBP to JKM is not at a negligible level, at 2.65%.

The spillover indexes of the volatility series are reported in Table 6. The total connectedness of the volatility series is 32.8%, while the total connectedness of the return series is 22.9%. Risks tend to be transmitted between markets rather than returns. This mutual relationship is similar to the tendency of the return series. However, the characteristic of volatility series is that the influence of HH on the other variables is larger. Additionally, we hardly observe the spillover effects from JKM to the other variables.

Table 5. Spillover index between return series.

	From				
To	Henry Hub (HH)	National Balancing Point (NBP)	Title Transfer Facility (TTF)	Japan/Korea Marker (JKM)	Others
HH	98.3	0.53	0.93	0.23	0.42
NBP	0.96	57.8	40.6	0.64	10.5
TTF	1.02	41.3	57.2	0.54	10.7
JKM	0.30	2.65	1.88	95.2	1.21
Others	0.57	11.1	10.8	0.35	22.9

Table 6. Spillover index between volatility series.

	From				
To	Henry Hub (HH)	National Balancing Point (NBP)	Title Transfer Facility (TTF)	Japan/Korea Marker (JKM)	Others
HH	93.3	2.51	4.01	0.24	1.69
NBP	12.9	39.3	47.4	0.36	15.2
TTF	21.8	19.5	58.4	0.30	10.4
JKM	5.38	5.52	11.1	78.0	5.49
Others	10.0	6.88	15.6	0.22	32.8

4.2. Spectral Analysis

The spectral analyses results of the return series spillover are reported in Table 7. The total connectedness from 1 to 5 business days, from 6 to 20 business day, and over 21 business days is 17.3%, 4.12%, and 1.49%, respectively. The short-term factors contribute most to the return spillover. The spillover effects of the return series are mostly explained by events within about one week. Arbitrage might contribute to these short-term spillover effects.

Table 7. Spectral analyses of spillover index between return series.

Bandwidth	To	From				
		Henry Hub (HH)	National Balancing Point (NBP)	Title Transfer Facility (TTF)	Japan/Korea Marker (JKM)	Others
1 to 5 business days	HH	82.1	0.39	0.72	0.18	0.32
	NBP	0.55	47.1	32.5	0.49	8.40
	TTF	0.54	31.2	45.8	0.37	8.02
	JKM	0.14	1.17	0.84	69.6	0.54
	Others	0.31	8.18	8.53	0.26	17.3
6 to 20 business days	HH	12.0	0.10	0.15	0.04	0.07
	NBP	0.30	7.94	5.93	0.11	1.59
	TTF	0.35	7.45	8.37	0.13	1.98
	JKM	0.11	1.05	0.74	18.4	0.47
	Others	0.19	2.15	1.71	0.07	4.12
21 business days onward	HH	4.29	0.04	0.05	0.01	0.03
	NBP	0.11	2.80	2.09	0.04	0.56
	TTF	0.13	2.66	2.97	0.05	0.71
	JKM	0.05	0.43	0.30	7.17	0.20
	Others	0.07	0.78	0.61	0.02	1.49

Table 8 presents the spectral analyses results of the volatility series spillovers. The total connectedness from 1 to 5 business days, from 6 to 20 business days, and over 21 business days is 0.10%, 2.37%, and 30.3%, respectively. Contrary to the return series, the long-term factors contribute most to the volatility spillover. Most of the spillover effect of the volatility series is caused by events that occurred more than one month ago. The reason might be the long-term memory of volatility.

Table 8. Spectral analyses of spillover index between volatility series.

Bandwidth	To	From				
		Henry Hub (HH)	National Balancing Point (NBP)	Title Transfer Facility (TTF)	Japan/Korea Marker (JKM)	Others
1 to 5 business days	HH	0.06	0.00	0.00	0.00	0.00
	NBP	0.03	0.50	0.17	0.00	0.05
	TTF	0.02	0.15	0.20	0.00	0.04
	JKM	0.01	0.02	0.02	22.0	0.01
	Others	0.01	0.04	0.05	0.00	0.10
6 to 20 business days	HH	2.09	0.04	0.01	0.00	0.01
	NBP	0.27	7.27	3.90	0.02	1.05
	TTF	0.20	3.63	4.10	0.02	0.96
	JKM	0.12	0.75	0.55	32.6	0.35
	Others	0.15	1.10	1.11	0.01	2.37
21 business days onward	HH	91.1	2.47	4.00	0.24	1.68
	NBP	12.7	31.6	43.3	0.33	14.1
	TTF	21.6	15.7	54.1	0.28	9.39
	JKM	5.26	4.75	10.5	23.5	5.13
	Others	9.87	5.74	14.5	0.21	30.3

4.3. Rolling Analysis

4.3.1. Total Connectedness

Figure 5 shows the total connectedness of the return series and the results of its spectral analyses. The total connectedness has not fluctuated significantly for 10 years and there is no significant change in the composition ratio of the total connectedness. The short-term factors have almost caused return spillovers.

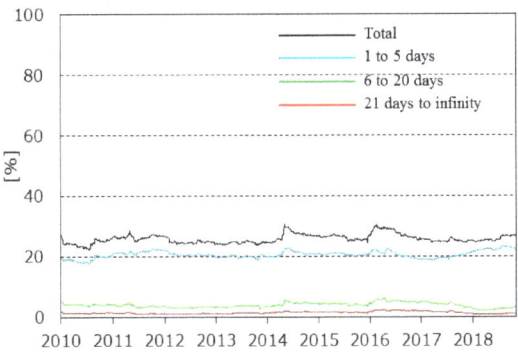

Figure 5. Total connectedness of return series.

Figure 6 represents the total connectedness of the volatility series and the results of its spectral analyses. As with the static analysis results, the long-term factors almost caused volatility spillovers.

Total connectedness spikes in February 2014, February 2018, and November 2018. We assume these spikes are caused by sudden fluctuations in each series. The price of HH spikes in February 2014 because of a great cold wave. The same phenomenon occurs every winter, although the scale is small. Additionally, the JKM, NBP, and TTF prices began to decline at around same time, because the spot market, which has been tight since the Great East Japan Earthquake, shifts and becomes loose. Therefore, the spillover index spikes.

The oversupply strengthens from January to February 2018, because of the expectation to expand USA's LNG export capacity along with the production of crude oil and natural gas. As a result, each gas price index falls sharply. Conversely, in November, the HH increases sharply because the storage stock levels in the USA fall significantly below recent levels. These facts cause the two spikes in 2018.

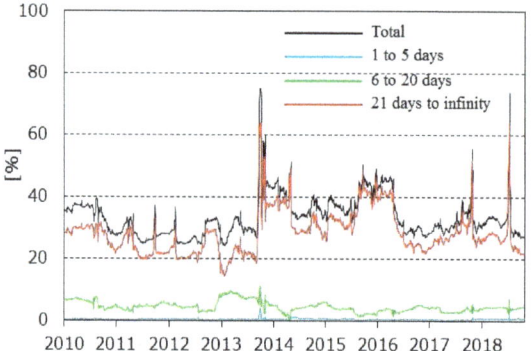

Figure 6. Total connectedness of volatility series.

4.3.2. Pairwise Connectedness

Figure 7 traces the pairwise connectedness of the return series and the results of its spectral analyses. In all combinations, the spillover effects of the return series continue to depend on short-term factors. Compared to the total connectedness, which is stable at around 30%, the pairwise connectedness between any variables is low and stable, although the pairwise connectedness from NBP to JKM and from TTF to JKM temporarily reaches around 3%. The mutual spillover effects offset each other because they are almost at the same level. However, when the Asia-Pacific spot market becomes tight or loose, the European market might be relatively stronger than the Asia-Pacific one.

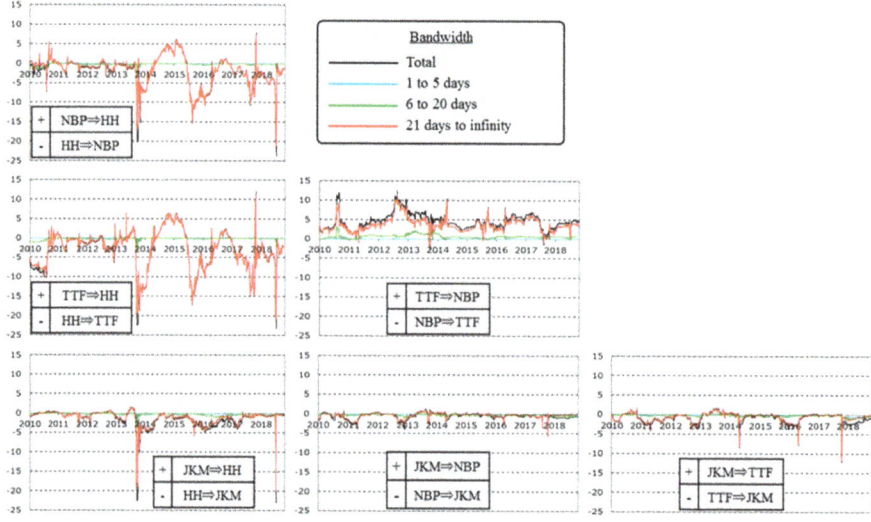

Figure 7. Pairwise connectedness of return series.

Figure 8 shows the pairwise connectedness of the volatility series and the results of its spectral analyses. Under all combinations, the spillover effects of the volatility series continue to depend on long-term factors. The spillover effects from HH to NBP and from TTF to JKM are periodically strong. The HH volatility spikes caused by the cold wave in North America affect the global market. In the European market, the spillover effect from TTF to NBP is stably high. TTF might be more prone to risks than NBP, because the former is more closely linked to intra-European and international trading using pipelines and LNG.

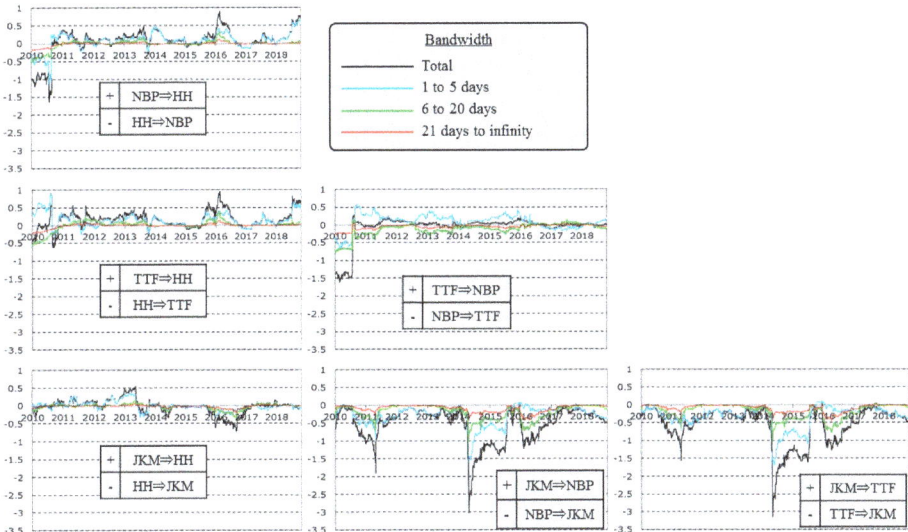

Figure 8. Pairwise connectedness of volatility series.

5. Conclusions

We adopt the approach of Diebold and Yilmaz [3] to examine the spillover effects between global natural gas price indexes. Moreover, we employ the Fourier transform utilized by Baruník and Křehlík [25] for spectrally decomposing Diebold and Yilmaz's [3] index.

We use daily data from 2 February 2009 to 28 February 2019. We employ HH, NBP, TTF, and JKM as the North American, British, Continental European, and Asia-Pacific market price indexes, respectively.

The results of the analyses for the entire period show that the total connectedness of return and volatility is 22.9% and 32.8%, respectively. Volatility is more highly integrated than returns in the global natural gas market. However, total connectedness is mostly dependent on the pairwise connectedness between NBP and TTF. Therefore, we cannot conclude that the global natural gas market is integrated.

The results of the spectral analyses indicate that the spillover effects of the return series almost depend on short-term factors, that is, events within 5 business days. If we further expand the natural gas pipeline network to produce more LNG, the total connectedness of the return series and the ratio of short-term factors should be higher with the increase in arbitrage trading. By contrast, the spillover effect of the volatility series is almost dependent on the long-term factors, that is, accumulated shocks for more than a month. This result represents the long-term memory of volatility.

The results of the dynamic analyses with moving window samples are consistent with the results above. The spillover effects of the return and volatility series tend to be dependent on events within a week and more than a month ago, respectively. However, regional climate, demand and supply, and incidents might make spillover effects unstable. Unfortunately, the spillover index for around 10 years does not show that the liquidity of the global natural gas market tends to increase.

For practitioners, this study implies that constantly monitoring market spillovers is significant. The information obtained by this study's approach is also of high value for energy portfolio rebalancing, including derivatives. Moreover, if we monitor the index as a predictive indicator, the information helps with early risk aversion.

Finally, clarification of the spillover effects between the crude oil and natural gas markets will be required for future studies, because many natural gas trading contracts still have their prices linked to crude oil prices.

Author Contributions: Data curation, T.N.; formal analysis, Y.T.; retry analysis, T.N.; writing, T.N.; editing, T.N.; supervision, T.N.

Funding: This research received no external funding.

Acknowledgments: The authors would like to thank the anonymous reviewers, whose valuable comments helped improve an earlier version of this paper.

Conflicts of Interest: The authors declare no conflict of interest.

References

1. BP's Statistical Review of World Energy. Available online: https://www.bp.com/en/global/corporate/energy-economics/statistical-review-of-world-energy.html (accessed on 29 September 2019).
2. Japan Fair Trade Commission; Government of Japan. Ekika-Ten-Nen-Gas-No-Torihiki-Jittai-Ni-Kansuru-Chousa-Houkokusho (Survey on LNG Trades). 2017. Available online: https://www.jftc.go.jp/houdou/pressrelease/h29/jun/170628_1_files/170628_4.pdf (accessed on 29 September 2019).
3. Diebold, F.X.; Yilmaz, K. Better to give than to receive: Predictive directional measurement of volatility spillovers. *Int. J. Forecast.* **2012**, *28*, 57–66. [CrossRef]
4. Kum, H.; Ocal, O.; Aslan, A. The relationship among natural gas energy consumption, capital and economic growth: Bootstrap-corrected causality tests from G-7 countries. *Renew. Sustain. Energy Rev.* **2012**, *16*, 2361–2365. [CrossRef]
5. Das, A.; McFarlane, A.A.; Chowdhury, M. The dynamics of natural gas consumption and GDP in Bangladesh. *Renew. Sustain. Energy Rev.* **2013**, *22*, 269–274. [CrossRef]
6. Bildirici, M.E.; Bakirtas, T. The relationship among oil, natural gas and coal consumption and economic growth in BRICTS (Brazil, Russian, India, China, Turkey and South Africa) countries. *Energy* **2014**, *65*, 134–144. [CrossRef]
7. Acaravci, A.; Ozturk, I.; Kandir, S.Y. Natural gas prices and stock prices: Evidence from EU-15 countries. *Econ. Model.* **2012**, *29*, 1646–1654. [CrossRef]
8. Nakajima, T.; Hamori, S. Testing causal relationships between wholesale electricity prices and primary energy prices. *Energy Policy* **2013**, *62*, 869–877. [CrossRef]
9. Atil, A.; Lahiani, A.; Nguyen, A.D. Asymmetric and nonlinear pass-through of crude oil prices to gasoline and natural gas prices. *Energy Policy* **2014**, *65*, 567–573. [CrossRef]
10. Perifanis, T.; Dagoumas, A. Price and Volatility Spillovers Between the US Crude Oil and Natural Gas Wholesale Markets. *Energies* **2018**, *11*, 2757. [CrossRef]
11. Tiwari, A.K.; Mukherjee, Z.; Gupta, R.; Balcilar, M. A wavelet analysis of the relationship between oil and natural gas prices. *Resour. Policy* **2019**, *60*, 118–124. [CrossRef]
12. Xia, T.; Ji, Q.; Geng, J.B. Nonlinear dependence and information spillover between electricity and fuel source markets: New evidence from a multi-scale analysis. *Physica A* **2020**, *537*, 122298. [CrossRef]
13. Batten, J.A.; Kinateder, H.; Szilagyi, P.G.; Wagner, N.F. Time-varying energy and stock market integration in Asia. *Energy Econ.* **2019**, *80*, 777–792. [CrossRef]
14. Nakajima, T. Expectations for Statistical Arbitrage in Energy Futures Markets. *J. Risk Financ. Manag.* **2019**, *12*, 14. [CrossRef]
15. Olsen, K.K.; Mjelde, J.W.; Bessler, D.A. Price formulation and the law of one price in internationally linked markets: An examination of the natural gas markets in the USA and Canada. *Ann. Reg. Sci.* **2015**, *54*, 117–142. [CrossRef]

16. Scarcioffolo, A.R.; Etienne, X.L. How connected are the U.S. regional natural gas markets in the post-deregulation era? Evidence from time-varying connectedness analysis. *J. Commod. Mark.* **2019**, *15*, 100076. [CrossRef]
17. Ren, X.; Lu, Z.; Cheng, C.; Shi, Y.; Shen, J. On dynamic linkages of the state natural gas markets in the USA: Evidence from an empirical spatio-temporal network quantile analysis. *Energy Econ.* **2019**, *80*, 234–245. [CrossRef]
18. Nick, S. The informational efficiency of European natural gas hubs: Price formation and intertemporal arbitrage. *Energy J.* **2016**, *37*, 1–30. [CrossRef]
19. Osička, J.; Lehotský, L.; Zapletalová, V.; Černoch, F.; Dančák, B. Natural gas market integration in the Visegrad 4 region: An example to follow or to avoid? *Energy Policy* **2018**, *112*, 184–197. [CrossRef]
20. Bastianin, A.; Galeotti, M.; Polo, M. Convergence of European natural gas prices. *Energy Econ.* **2019**, *81*, 793–811. [CrossRef]
21. Shi, X.; Shen, Y.; Wu, Y. Energy market financialization: Empirical evidence and implications from East Asian LNG markets. *Financ. Res. Lett.* **2019**, *30*, 414–419. [CrossRef]
22. Neumann, A. World natural gas markets and trade: A multi-modeling perspective. *Energy J.* **2009**, *30*, 187–199. [CrossRef]
23. Chai, J.; Wei, Z.; Hu, Y.; Su, S.; Zhang, Z.G. Is China's natural gas market globally connected? *Energy Policy* **2019**, *132*, 940–949. [CrossRef]
24. Silverstovs, B.; L'Hégaret, G.; Neumann, A.; Hirschhausen, C. International market integration for natural gas? A cointegration analysis of prices in Europe, North America and Japan. *Energy Econ.* **2005**, *27*, 603–615. [CrossRef]
25. Baruník, J.; Křehlík, T. Measuring the frequency dynamics of financial connectedness and systemic risk. *J. Financ. Econom.* **2018**, *16*, 271–296. [CrossRef]
26. Toyoshima, Y.; Hamori, S. Measuring the time-frequency dynamics of return and volatility connectedness in global crude oil markets. *Energies* **2018**, *11*, 2893. [CrossRef]
27. Ji, Q.; Geng, J.B.; Tiwary, A.K. Information spillovers and connectedness networks in the oil and gas markets. *Energy Econ.* **2018**, *75*, 71–84. [CrossRef]
28. Engle, R.F.; Kroner, K.F. Multivariate Simultaneous Generalized ARCH. *Econom. Theory* **1995**, *11*, 122–150. [CrossRef]

© 2019 by the authors. Licensee MDPI, Basel, Switzerland. This article is an open access article distributed under the terms and conditions of the Creative Commons Attribution (CC BY) license (http://creativecommons.org/licenses/by/4.0/).

MDPI
St. Alban-Anlage 66
4052 Basel
Switzerland
Tel. +41 61 683 77 34
Fax +41 61 302 89 18
www.mdpi.com

Energies Editorial Office
E-mail: energies@mdpi.com
www.mdpi.com/journal/energies

www.ingramcontent.com/pod-product-compliance
Lightning Source LLC
LaVergne TN
LVHW070156120526
838202LV00013BA/1149